SpringerBriefs in Applied Sciences and Technology

SpringerBriefs present concise summaries of cutting-edge research and practical applications across a wide spectrum of fields. Featuring compact volumes of 50 to 125 pages, the series covers a range of content from professional to academic.

Typical publications can be:

- A timely report of state-of-the art methods
- An introduction to or a manual for the application of mathematical or computer techniques
- A bridge between new research results, as published in journal articles
- A snapshot of a hot or emerging topic
- An in-depth case study
- A presentation of core concepts that students must understand in order to make independent contributions

SpringerBriefs are characterized by fast, global electronic dissemination, standard publishing contracts, standardized manuscript preparation and formatting guidelines, and expedited production schedules.

On the one hand, **SpringerBriefs in Applied Sciences and Technology** are devoted to the publication of fundamentals and applications within the different classical engineering disciplines as well as in interdisciplinary fields that recently emerged between these areas. On the other hand, as the boundary separating fundamental research and applied technology is more and more dissolving, this series is particularly open to trans-disciplinary topics between fundamental science and engineering.

Indexed by EI-Compendex, SCOPUS and Springerlink.

Tin-Chih Toly Chen

Explainable Artificial Intelligence in Supply Chain Management

Methodology, System Architecture, and Applications

Springer

Tin-Chih Toly Chen
Department of Industrial Engineering
and Management
National Yang Ming Chiao Tung University
Hsinchu, Taiwan

ISSN 2191-530X ISSN 2191-5318 (electronic)
SpringerBriefs in Applied Sciences and Technology
ISBN 978-3-032-22563-4 ISBN 978-3-032-22564-1 (eBook)
https://doi.org/10.1007/978-3-032-22564-1

This Springer imprint is published by the registered company Springer Nature Switzerland AG
The registered company address is: Gewerbestrasse 11, 6330 Cham, Switzerland

Competing Interests The author has no competing interests to declare that are relevant to the content of this manuscript.

Contents

Chapter 1
Explainable Artificial Intelligence (XAI) Applications in Supply Chain Management (SCM)

1.1 Artificial Intelligence (AI) Applications in Supply Chain Management (SCM)

Artificial intelligence (AI) technologies, such as machine learning (ML) and deep learning (DL) technologies, have been widely used to help humans or robots solve reasoning, prediction, optimization, and classification problems in various fields [1].

A **supply chain**, also called a demand chain (from the customer perspective), supply network, value chain, or value network, represents the complex networks and links of businesses entities to make and deliver a product and/or service to customers, as illustrated in Fig. 1.1, in which the focal firm is the key product and/or service provider within the supply chain. The focal firm may not actually make any physical goods or delivery service itself, e.g., Apple Inc. in the supply chain of iPhone. However, without the focal firm, a supply chain would not exist.

Supply chain management (SCM) is the systematic and strategic integration and coordination of various business functions and processes across the supply chain to create value for the ultimate customers in an effective and efficient manner [2]. To this end, a lot of AI applications can already be seen. Liu [2] divided the activities of SCM into six categories: procurement, inbound logistics, production, outbound logistics, order fulfillment, and after-sale services.

Before AI became as widespread as it is today, according to Min [3], the main AI methods used in SCM include artificial neural networks (ANNs), rough set theory, ML, expert systems, genetic algorithms (GAs), fuzzy logic, and agent-based systems. However, both ANNs and GAs belong to ML [1]. In addition, state-of-the-art AI technologies, such as DL [4], explainable artificial intelligence (XAI) [5], and generative artificial intelligence (GenAI) [6], are not involved.

Another recent review by Daios et al. [7] summarized the applications of AI in supporting various functionalities of SCM, including customer relationship management (CRM), inventory management, transportation networking, procurement, demand forecasting, resilience enhancement, and risk management, as follows:

T.-C. T. Chen, *Explainable Artificial Intelligence in Supply Chain Management*, SpringerBriefs in Applied Sciences and Technology,
https://doi.org/10.1007/978-3-032-22564-1_1

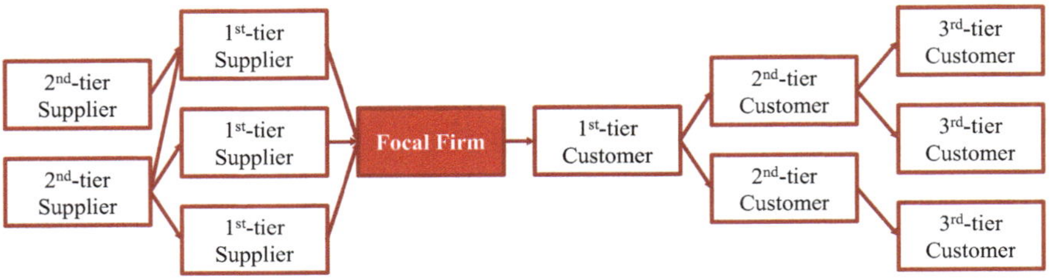

Fig. 1.1 Supply chain

- CRM: Agent-based models, chatbots, and virtual assistants.
- Inventory management: ML, robots, drones, agent-based models, and large language models (LLMs).
- Transportation networking: Network theory, graph algorithms, GAs, ant colony optimization, and reinforcement learning.
- Procurement: Agent-based models, process automation, and GenAI.
- Demand forecasting: ML, support vector machines (SVMs), ANNs, decision trees, deep neural networks (DNNs), data mining, fuzzy logic, sentiment analysis, and LLMs.
- Resilience enhancement: ANNs, deep reinforcement learning, and Bayesian networks.
- Risk management: Ensemble learning, ANNs, fuzzy optimization, ML, big data, agent-based systems, GenAI, DNNs, and LLMs.

This can be summarized as 26 common AI applications: agent-based models, ANNs, ant colony optimization, Bayesian networks, big data, chatbots, data mining, decision trees, deep reinforcement learning, DNNs, drones, ensemble learning, fuzzy logic, fuzzy optimization, GenAI, GAs, graph algorithms, LLMs, ML, network theory, process automation, robots, reinforcement learning, sentiment analysis, SVMs, and virtual assistants.

It is worth noting that the SCM functionalities supported by AI applications mentioned by Daios et al. [7] do not cover all six task categories defined by Liu [1], as shown in Fig. 1.2, which means that some SCM tasks may have received very little support from AI applications so far.

In addition, SCM relies heavily on the communication and collaboration between upstream and downstream participants via networked information systems. Such communication and collaboration are often lengthy and have tactical or strategic significance. AI technologies excel at predicting/estimating, controlling, classifying, and optimizing in short-term activities during the execution phase, but are less suitable for such cross-organizational managerial tasks.

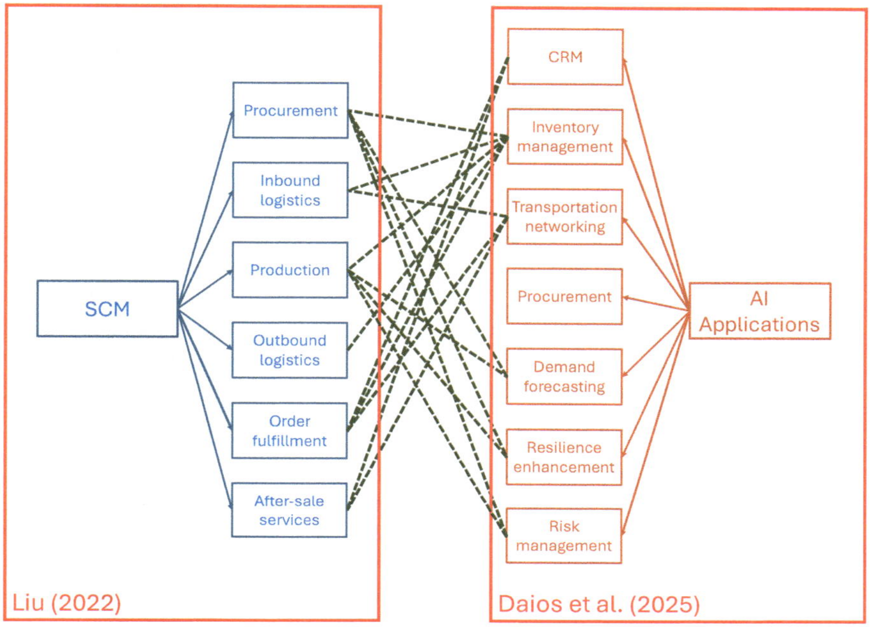

Fig. 1.2 SCM tasks support by AI applications

1.2 Explainable Artificial Intelligence (XAI)

Some AI technologies are overly complex or opaque, making it difficult for users to understand, communicate, or trust them, thus affecting their willingness and frequency of application. To address this issue, **XAI** emerged as another noteworthy trend in the development of AI [8]. In recent years, GenAI services have become increasingly popular, enabling users to communicate with high-performance servers in natural language to meet various needs and achieve similar goals to XAI [9, 10].

According to Defense Advanced Research Projects Agency (DARPA) [10], XAI is to generate more explainable ML models, while maintaining a high level of reasoning performance (prediction, classification accuracy, or control precision). XAI enables humans to understand, appropriately trust, and effectively manage these AI applications or their agents. General Data Protection Regulation (GDPR) in European countries defines the concept of XAI as customers having the right to interpret decisions made through automated systems [11]. However, what kinds of AI technologies are explainable is inconclusive. In the view of McNamara [12], ensemble methods, random forecasts, decision trees, Bayesian networks, sparse linear models, and others are highly explainable, while ANNs, DL, type-*n* fuzzy logic, SVMs, and others are less explainable. Therefore, XAI techniques and tools are referred to as white boxes or the twins of AI applications that are often considered black boxes [13].

In most black-box AI applications, it is difficult to infer the result from the input data. In SCM, mathematical programming (MP) models are typically built and solved to optimize supply chain performance. Some MP models, such as nonlinear programming (NLP) models, are difficult to handle, meaning it is hard to find the globally optimal solution. However, it is not impossible to infer the optimal solution from the input data. For example, we already know that some possible solutions are good, but we need to check whether they are feasible or whether they can be fine-tuned to make them better. Therefore, whether such MP models belong to the category of black-box problems remains controversial. In addition, some problem-solving techniques, such as branch and bound (B&B), GAs [14], and ant colony optimization (ACO) [15], are also used to assist in solving MP problems. These problem-solving techniques are often black boxes which become the targets of explanations (see Fig. 1.3). However, few studies have attempted to explain them.

So far, XAI has developed in at least three directions (see Fig. 1.4):

- XAI enhances the practicality of an existing AI technology by explaining its reasoning process and/or result [10]. For example, local interpretable model-agnostic explanation (LIME) is to explain the classification or prediction result using an ML (or DL) model, such as a DNN, by identifying critical features/predictors and fitting simple and locally interpretable models, such as linear

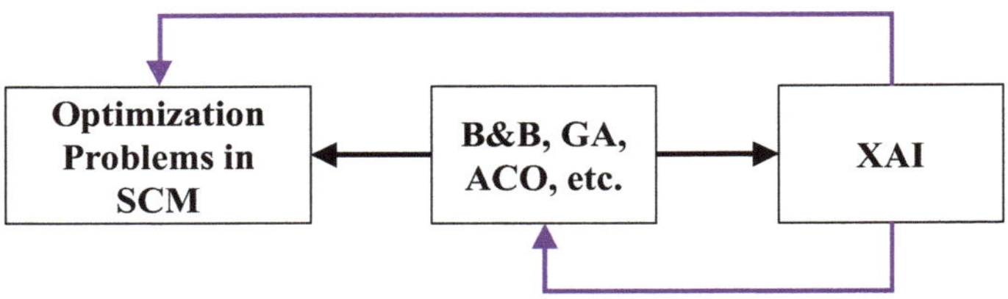

Fig. 1.3 XAI for explaining applications of problem-solving techniques in SCM

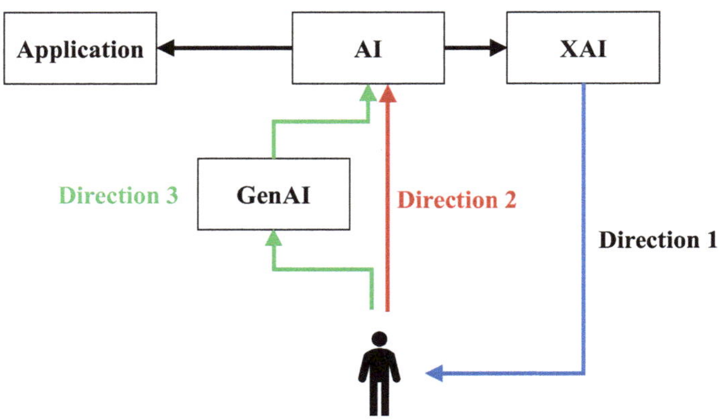

Fig. 1.4 Directions of XAI development

regression or decision trees [16]. In LIME, a synthetic data set is generated and fed into the ML (or DL) model to generate their outputs. Simple decision rules or linear regression equations are then fitted to explain the reasoning mechanism of these synthetic data [17, 18]. Some studies have used heatmaps to emphasize the parts of an image that a DNN emphasizes when classifying or recognizing patterns, which is similar [19, 20].

- XAI is to improve the effectiveness of existing AI technologies by incorporating easy-to-interpret visual features, such as heatmaps [19, 20], decision (or regression) rules [21], decision trees [18, 21], and scatter plots [22], to help diagnose their reasoning mechanisms. An example is attention-based DNNs that incorporate DNNs with heatmaps [19, 20]. However, the linkage between visual features and the reasoning mechanism needs to be discovered.
- XAI is to enhance the convenience and effectiveness of AI applications by using natural language processing (NLP), DL, and accelerated computing, namely, GenAI [6, 9], to make user queries more friendly and collect and reorganize Web data faster to fulfill specific tasks.

1.3 Implementation Procedure of XAI

According to Chen [8], the implementation procedure of XAI comprises seven steps, as illustrated in Fig. 1.5. For an AI technology application, the reasoning process and/or result should be explainable and interpretable. In addition, the result needs to be validated [23]. After these are done, the AI technology application can be said to be explainable, which contributes to its trustworthiness, fairness, accountability, and transparency [8, 16, 17]. In addition, the explainability of AI technologies fosters improvement ideas based on which these AI technologies can be modified to enhance their effectiveness [24, 25].

In SCM, many AI applications involve the formulation and solving of optimization problems, such as production, logistics planning, and job scheduling. Fairness of such optimization models is usually satisfactory, since anyone can easily check the feasibility and/or optimality of any possible solution. Transparency is also good since the reasoning mechanism is visible, i.e., the objectives and constraints. Traditionally, sensitivity or parametric analyses are usually conducted to provide operational insights into the optimization problems, which is beneficial to.

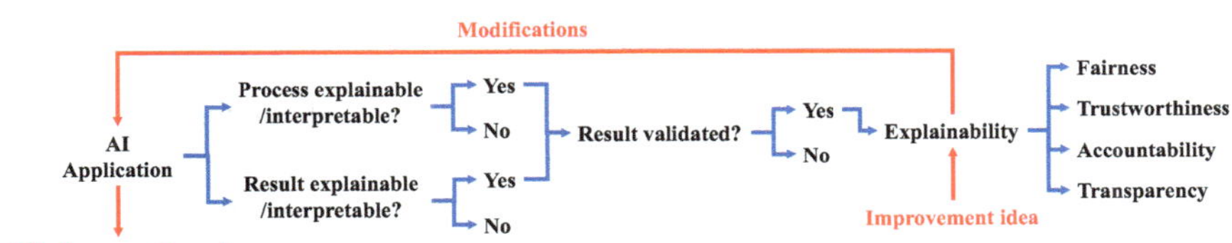

Fig. 1.5 Implementation procedure of XAI [8]

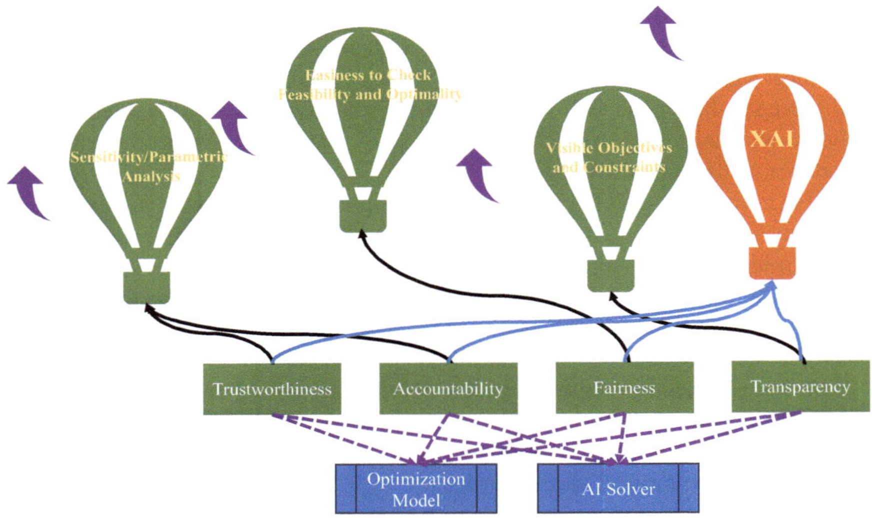

Fig. 1.6 Traditional methods for explaining the formulation and solving of optimization problems in SCM

- Trustworthiness: Supply chain planners can check the results against their domain knowledge.
- Accountability: Supply chain planners can consider any advantages or disadvantages that may arise from various possible reasons.

This is illustrated in Fig. 1.6. These phenomena may decrease the motivation for explaining the AI applications in helping solve the optimization problems for SCM.

Another mainstream of AI applications in SCM is multi-criteria decision-making (MCDM) or group decision-making (GDM) [26], such as facility location selection, supplier selection, and evaluation. With more sophisticated decision-making models and more diverse experts' judgments and evaluations, it becomes challenging to infer the final decision-making result from the input data, requiring the intervention of XAI techniques or tools.

1.4 XAI Applications in SCM

Some of the AI methods used in SCM are difficult to comprehend or communicate, such as DNNs, SVMs, and GAs, the so-called black-box AI models [27], while others are easier to understand and use, such as decision trees and fuzzy logic [28]. Theoretically, XAI techniques and tools can be applied to explain the reasoning process and result of a black-box AI application. However, so far, such attempts have mostly focused on a few AI applications, as shown in Fig. 1.7.

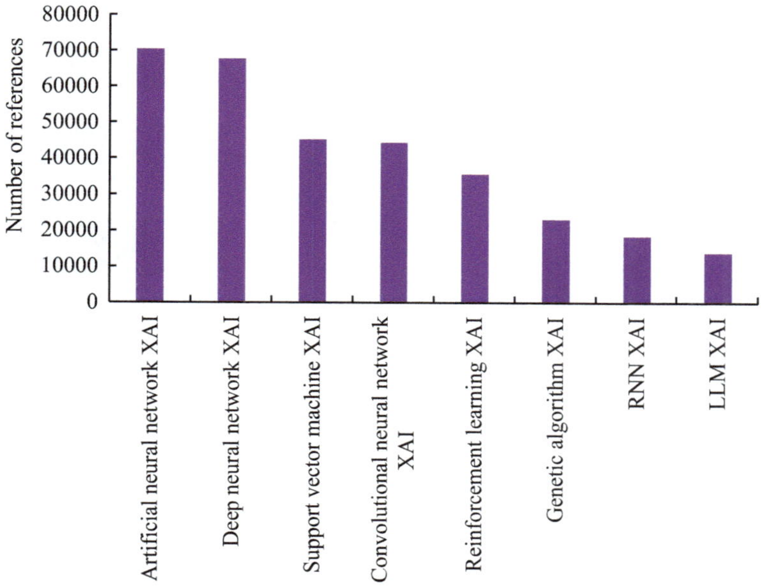

Fig. 1.7 XAI applications to explain black-box models. *Data source* Google Scholar, until 2026/1/6

On the one hand, if there have been sufficient attempts to explain AI applications in one functionality of SCM, then these AI applications will become more transparent and credible. On the other hand, this phenomenon exists just because these AI applications have low transparency and credibility. Ultimately, it seems that a balance has been reached regarding the transparency of AI applications across all SCM functionalities.

Mugurusi and Oluka [29] categorized the applications of XAI in supply chain management into four types: the first type is used for tracking and reasoning (how and why decisions are made); the second type is used for argumentation or support (explaining the logic behind each decision); the third type is used to explain system control behavior and problem-solving strategies; and the fourth type is used to provide terminology details. This classification ranges from the simplest to the most complex, from operational details to strategies, but lacks support from real-world examples.

Some examples of XAI applications in SCM are reviewed as follows. These examples can be divided into the following categories:

- XAI applications in product quality management within supply chains.
- XAI applications in supply chain resilience.
- XAI applications in demand forecasting for supply chains.
- XAI applications in supplier selection for supply chains.
- XAI applications in logistics management for supply chains.
- XAI applications in supply chain performance evaluation [26].

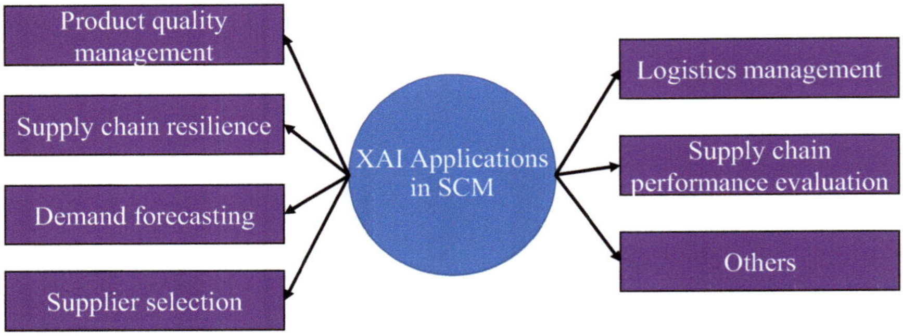

Fig. 1.8 XAI applications in various topics of SCM

This is shown in Fig. 1.8.

1.4.1 XAI Applications in Product Quality Management Within Supply Chains

For a food supply chain, transmitting complete information about food ingredients from the upstream to the downstream without tampering is a crucial task. Bhatia and Albarrak [30] designed an online food safety testing system. Users could scan the bar code of a food product to obtain its ingredient information, which was then fed into a region-based convolutional neural network (RCNN) to determine whether the food was safe. They also applied XAI techniques, such as Shapley value (SHAP) analysis, to identify the most important ingredient features, thereby improving the classification performance of the RCNN.

In multi-tiered supply chains, the causes of changes in product quality and delivery durations are difficult to identify and can only be partially observed because these causes are usually located outside the company. To address this issue, traditional SCM advocates for data sharing to gain deeper insights, but this practice has not been implemented in practice due to concerns about data privacy. To address this issue, Schoepf and Foster [31] applied XAI techniques, especially SHAP analysis, to perform decentralized calculations for estimating product quality variation using ANNs in multi-stage production processes, thereby reducing the need to persuade supply chain participants to share data, since all calculations were performed in a decentralized manner. In traditional centralized systems, the impacts of inputs to the ANN were evaluated with the corresponding SHAP values, and then each participant's contribution was assessed by summing the SHAP values of all inputs provided by the participant. In contrast, in decentralized systems, each participant provided only their own estimates, and then another ANN was built to aggregate the estimates from all participants.

1.4.2 XAI Applications in Supply Chain Resilience

Sadeghi et al. [32] argued that XAI applications could help close the gap between other firms and pioneering firms in an industry in terms of the effectiveness of using AI-driven technologies. Furthermore, they conducted an experiment in which participants were divided into two groups and asked to answer questions. One group used AI technologies (audio and images) to assist them, while the other group also applied XAI tools, such as voice AI assistants, to help them understand AI applications. They then compared the performances of the two groups. The comparison results were used to test the hypothesis that whether XAI could enhance the resilience of supply chain networks.

1.4.3 XAI Applications in Demand Forecasting for Supply Chains

In the development of a perishable goods supply chain, optimizing costs and profits while meeting customer needs is a key challenge. Jauhar et al. [33] believed that customer-centric demand forecasting could address this challenge by considering customer characteristics when determining inventory levels. They proposed a two-step approach: first, segmenting based on customer characteristics using k-means and hierarchical clustering, and then forecasting demand for each segment. The results showed that hierarchical clustering and long short-term memory network (LSTM) performed best in demand forecasting. Subsequently, to improve the transparency of the forecasting model, they built a surrogate model in the form of random forest (RF). However, the interpretability of random forests is actually limited due to the large number of decision trees and the potential for user confusion caused by related explanations.

The COVID-19 pandemic has severely impacted various supply chains. There was a general labor shortage in all links of a supply chain. Alhazmi et al. [34] constructed LSTMs to predict the number of confirmed cases and recovery status of workers in each link of a supply chain. They then used SHAP analysis to assess the importance of each input variable to the prediction result. However, they pointed out that the importance of some input variables may not be constant during the long COVID-19 pandemic. Therefore, they believed that the changes in the importance of an input variable could be traced through time series models, thereby adjusting the future importance of the input variable and making more accurate decisions.

1.4.4 XAI Applications in Supplier Selection for Supply Chains

Choosing a suitable site to meet customer capacity requirements is an emerging issue in semiconductor supply chain management. However, this topic has not been fully studied. To this end, Chen et al. [35] proposed a fuzzy group decision-making (FGDM) method based on XAI applications to assist wafer foundries in choosing a suitable site, in which a variety of visualization tools and methods were introduced, including hanging gradient bar charts, gradient bidirectional scatterplots, and hanging gradient bar charts for traceable aggregation. These new XAI tools significantly improved the interpretability of the site selection process and result, with an advantage of up to 36% compared to existing XAI tools. In addition, SHAP analysis results revealed that factors having the greatest impact on the evaluation result might differ from the original judgments of domain experts.

Ubiquitous manufacturing is characterized by supply chain networks with unlimited capacity and perpetually available manufacturing services [36]. Wang and Chen [37] proposed four new XAI techniques and tools for selecting appropriate 3D printing facilities in ubiquitous manufacturing networks: gradient bar charts with baseline, group gradient bar charts, manually adjustable gradient bar charts, and bidirectional scatterplots. These XAI tools support four goals of XAI: understandability, comprehensibility, interpretability, and transparency [38]. They also listed 12 requirements for evaluating the effectiveness of XAI applications in ubiquitous manufacturing networks.

According to Chen [39], the results of cycle time prediction, analysis, and modeling can support various management activities, including SCM. Wang et al. [40] constructed a DNN to accurately predict the cycle times of jobs in wafer fabrication plant (wafer fab). However, the prediction mechanism of the DNN was complex and difficult to interpret. To address this issue, they proposed an interpretable DL approach, in which a classification regression tree (CART) was first built to explain the job cycle time prediction mechanism of the DNN. Then, the predicted value of each branch in CART was replaced with a fuzzy linear regression (FLR) equation, which was used to estimate the cycle time range to compensate for the lack of interpretability.

1.4.5 XAI Applications in Logistics Management for Supply Chains

Ahmed et al. [41] proposed a hybrid DL approach that combines self-organizing map (SOM), principal component analysis (PCA), and ANN to predict shipping times, thereby assessing the likelihood of delayed shipments showing supply chain delivery risks. They also conducted SHAP analysis to understand how each feature affected the prediction result. Experimental results showed that the mode of transport, order date, and shipment date were the most critical features for predicting shipping times.

In a supply chain, although consumer sales appear to fluctuate little, the volume of orders placed by retailers with wholesalers fluctuates significantly. In addition, the volume of orders placed by wholesalers with manufacturers and by manufacturers with suppliers varies even more over time. The bullwhip effect suggests that as the supply chain extends upstream from retail to manufacturing, the volatility of orders also increases. Many studies have applied AI technologies to more accurately predict demand in order to mitigate the effects of the bullwhip effect, such as the discrete wavelet transform (DWT) analysis and ANN method, used in Jaipuria and Mahapatra [42]. In the literature, ANNs and DNNs, compared with other AI technologies, have been extensively explained by many XAI studies [5, 8, 17–21].

1.4.6 XAI Applications in Supply Chain Performance Evaluation

Saen et al. [43] constructed a network model of a sustainable supply chain for the oil industry, which contained recurring links, where outputs from some stages re-entered the previous stages as inputs. Therefore, outputs should be minimized while maximizing output utilization. Subsequently, they extended the designed network model to a dynamic data envelope analysis (DEA) model. Finally, they used an ANN to predict the performance of the supply chain. To improve the interpretability of the method, they approximated the nonlinear operations in the network model with linear ones, similar to the concept of local interpretable model-agnostic explanations (LIME) [44], and performed some parameter analysis to examine the impact of key features.

Fuzzy inference systems (FISs) have been widely applied to supply chain performance evaluation. For example, Pourjavad and Shahin [45] constructed a Mamdani FIS to evaluate the green supply chain management performances of four industrial and household paint manufacturers in Iran. Ganga and Carpinetti [46] developed a Mamdani FIS to evaluate supply chain performances based on level 1 and level 2 performance indicators of the supply chain operations reference model (SCOR). However, although linguistic terms and fuzzy inference rules (FIRs) can be easily comprehended, it is still difficult to infer the output of an FIR from the inputs, making the FIS a black box.

XAI techniques applied in supply chain management are summarized in Table 1.1.

Table 1.1 XAI techniques applied in supply chain management

References	Supply chain type	Purpose	AI technologies	XAI techniques
Bhatia and Albarrak [30]	Food supply chain	Food safety testing	RCNN	SHAP
Sadeghi et al. [32]	Not restricted	Supply chain resilience	Audio and images	Voice AI assistant
Jauhar et al. [33]	Perishable goods supply chain	Demand forecasting	• k-means • Hierarchical clustering • LSTM	RF
Chen et al. [35]	Semiconductor supply chain	Suitable site selection	FGDM	• Hanging gradient bar charts • Gradient bidirectional scatterplots • Hanging gradient bar charts for traceable aggregation • SHAP
Wang and Chen [37]	UM system	Appropriate 3D printing facilities selection	• Alpha cut operation • Fuzzy intersection • Fuzzy technique for order preference by similarity to ideal solution	• Gradient bar charts with baseline • Group gradient bar charts • Manually adjustable gradient bar charts • Bidirectional scatterplots
Schoepf and Foster [31]	Multiechelon supply chain	Product quality variation estimation	ANN	SHAP
Alhazmi et al. [34]	Supply chains during the COVID-19 pandemic	Confirmed and recovered cases prediction	LSTM	SHAP
Ahmed et al. [38]	NA	Shipping time prediction	SOM, PCA, ANN	SHAP
Saen et al. [41]	Oil supply chain	Supply chain performance prediction	DEA, ANN	Linear approximation

1.5 Difficulties in Applying XAI in SCM

So far, XAI has had many applications in SCM. However,

- Supply chain participants typically operate primarily through networked information systems, rarely applying AI methods or seeking to understand the reasoning processes behind AI models. Most research findings originate from theoretical discussions by researchers. In other words, the motivation for practical application is not so obvious.
- Most AI applications in supply chains still focus on short-term operational tasks, which is not much different from AI applications in a single factory.
- Like AI applications in other fields, XAI applications in SCM are currently limited to a few functions, such as more mature interpretation methods like SHAP and LIME.
- Applying the results obtained using XAI applications is a challenge. For example, using SHAP analysis to identify the factor most critical to supply chain resilience does not mean that resilience can be improved by proportionately pursuing better performance of the critical factor. In XAI techniques, using partial derivative (PD) to evaluate the importance of factors can meet this need more accurately [47]. This challenge becomes even more daunting in supply chains, where cross-organizational collaboration may be required.
- Supply chain participants are applying XAI techniques on their own, which not only improve their own performances but also contributes to the entire supply chain. Therefore, it is difficult to say whether these XAI applications are internal factory applications or supply chain applications.
- Typical supply chain operations involve collaboration among many companies, yet reports on XAI applications are relatively scarce. Most cases focus on single-company XAI applications.
- Many supply chain planning problems are modeled as complex optimization problems and bio-inspired algorithms such as GA and ACO are employed to assist in solving them. Although the applications of these bio-inspired algorithms also fall under the category of black-box AI, they are rarely the focus of explanation. Most efforts have been made to explain ANN or DNN applications [5, 8, 17–21].
- An AI application may show great effect on one participating company, but unless the participating company is a bottleneck in the supply chain, it may not have an impact on the performance of the entire supply chain, thus reducing the incentive to interpret the AI application.

1.6 Basic XAI Concepts, Techniques, and Tools

Some basic XAI concepts, techniques, and tools are introduced as follows:

- A **local explanation** is to explain the rationale behind the result of a single example. The most widely used XAI technique for generating local explanations is SHAP, which evaluates the importance of each input attribute to the output. State-of-the-art XAI methods not only evaluate the importance of each input attribute to the output but also its importance to the parameters in the black-box model. This allows for the removal of less important parameters, making the black-box model more transparent or effective.
- A **global explanation** is to explain how the model arrives at its prediction/result and can be in the form of visualization (such as system diagrams and flowcharts), mathematical formulae, pseudo-code, performance measures, etc.
- **Contrastive explanations** help us understand why a model generates a certain result/output for a given input and not another: contrastive explanations also help determine the minimal changes in inputs or model parameters required to generate a different output. Correlation matrixes can also be used to show contrastive explanations, as illustrated in Fig. 1.9.
- **What-if explanations** are probably the most useful explanation technique, which helps us understand the relationship between the model result and input features, similar to the concept of a parametric analysis.
- **Counterfactual explanations** tell us how changing assumptions about the inputs or parameters of a model can cause the model to arrive at a particular different result, similar to the concept of a sensitivity analysis.

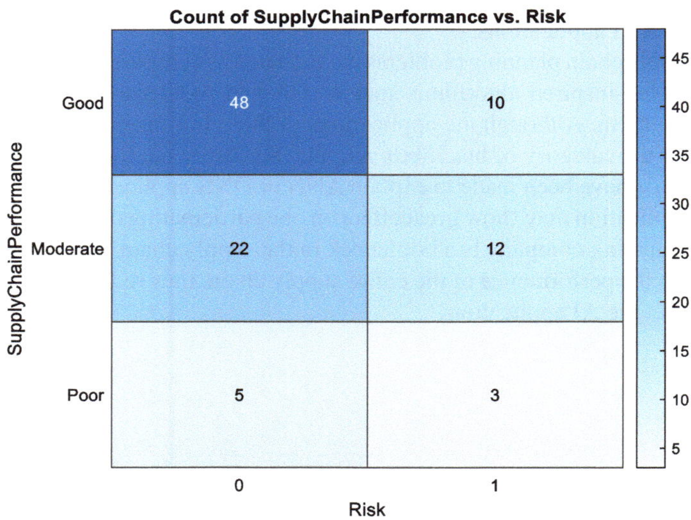

Fig. 1.9 Correlation matrix for showing contrastive explanation

- **Example-based explanations** are the simplest explanations in which the behavior of a model or underlying data distribution is simply explained by highlighting specific instances of the data.
- **AnchorTabular explanations** are similar to contingency table analyses, but perform perturbation.
- **Deep learning important features** (DeepLIFT) is a method for evaluating the importance scores of input features in DNNs. DeepLIFT works by comparing the activation of each neuron in the DNN giving the actual input with the activation generated giving a reference input, and then using the difference between the two to evaluate the importance scores of input features.
- **TreeSHAP**: TreeSHAP is an algorithm to compute SHAP values for decision tree-based models, such as CARTs, RFs, and XGBoost.

Methods for generating various explanations are summarized in Table 1.2.

Table 1.2 Methods for generating various explanations

Type of explanations	Methods
Local explanations	• LIME • SHAP analysis • Etc
Global explanations	• System diagrams • Flowcharts • Mathematical formulae • Pseudo-codes • Performance measures – Forecasting: mean absolute error (MAE), mean absolute percentage error (MAPE), root mean squared error (RMSE), R^2 – Classification: Recall, precision, F1, true positive rate (TPR), false positive rare (FPR), receiver operating characteristic (ROC) curve, area under curve (AUC), hit rate, cross-entropy loss – Decision-making: objective function value • Etc.
Contrastive explanations (or counterfactual explanations)	• Contingency table analysis • Cross-tabulation • Local foil tree method • Etc
What-if explanations	• Decision rules • Sensitivity analysis • Parametric analysis • Etc
Example-based explanations	• CBR • kM • Fuzzy c-means • Etc

A **model-agnostic XAI** method is that the computation required to explain an AI technology application is not based on the parameters of the AI technology application. **A post hoc interpretability method** explains the result, while **a pre hoc interpretability method** explains the process. Kamath and Liu [16] reviewed statistical and data analysis methods and tools suitable for the explanation of AI applications. In particular, they divided DL explanation methods into three categories: intrinsic methods, perturbation methods, and gradient/backpropagation methods.

Venugopal et al. [18] define the concept of **explainability failure** as a situation in which the output of an AI technology application is correct, but cannot be explained.

1.6.1 LIME

LIME is one of the most popular XAI techniques. LIME minimizes the following objective function [47]:

$$\text{Min } \xi(x) = \arg\min_{g \in G} \mathcal{L}(f, g, \pi_x) + \Omega(g) \tag{1.1}$$

where

f	the original model,
g	the interpretation model,
π_x	the proximity from all approximate outputs to the original output,
$\arg\min_{g \in G} \mathcal{L}()$	returns the g value that minimizes $\mathcal{L}()$, and
$\Omega()$	is a measure of model complexity.

LIME is implemented according to the following steps:

- Pre-model data analysis, using techniques such as bar-and-whisker plots, density plots, violin plots (for verifying data distribution), correlation analyses, correlation matrixes, heatmaps (for discovering data relationships), and backward regression elimination (for identifying most important inputs).
- AI model training.
- Synthetic data generation: The most common way to generate synthetic data is randomization. For inputs with correlation, synthetic data can be generated by randomizing the major input before deriving the other. Professional expertise is important for generating synthetic data for special cases such as medical applications.
- AI application to synthetic data.
- Approximate AI application by fitting synthetic data: Although most previous studies have constructed linear regression (LR) equations to fit the synthetic data in local feasible regions, some constructed decision rules to approximate local relationships with stepwise values.
- Explanation performance evaluation.

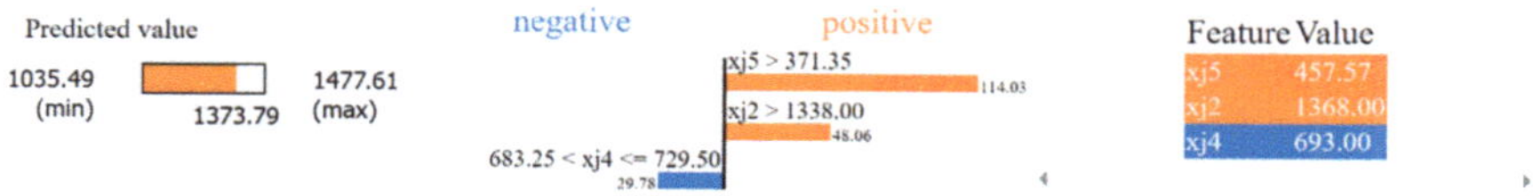

Fig. 1.10 Decision rules for approximating local relationships with stepwise values

LIME results are usually presented using LIME summary plots, as illustrated in Fig. 1.10. The required Python code is provided in Fig. 1.11.

```
# -*- coding: utf-8 -*-
# pip install lime
import lime
import lime.lime_tabular
import pandas as pd
import numpy as np
from sklearn.model_selection import train_test_split
from sklearn.metrics import mean_squared_error, mean_absolute_error,
mean_absolute_percentage_error, r2_score
from sklearn.neural_network import MLPRegressor

file_path = "F:\\XAI-class\\cycle time forecasting example.xlsx"
df = pd.read_excel(file_path, sheet_name="工作表1")
X = df[["xj1", "xj2", "xj3", "xj4", "xj5", "xj6"]]
y = df["aj"]


feature_names_list = X.columns.tolist()

X_train, X_test, y_train, y_test = train_test_split(X, y, test_size=0.2,
random_state=42)
mlp = MLPRegressor(random_state=42, max_iter=500)
mlp.fit(X_train.values, y_train)
y_pred = mlp.predict(X_test.values)
r2 = r2_score(y_test, y_pred)
rmse = np.sqrt(mean_squared_error(y_test, y_pred))
mape = mean_absolute_percentage_error(y_test, y_pred)
mae = mean_absolute_error(y_test, y_pred)
print("=== Model Performance ===")
print(f"R2   : {r2:.4f}")
print(f"RMSE : {rmse:.4f}")
print(f"MAPE : {mape:.4f}")
print(f"MAE  : {mae:.4f}")

explainer = lime.lime_tabular.LimeTabularExplainer(
    training_data=X_train.values,
    feature_names=feature_names_list,
    class_names=['regression_output'],
    mode='regression'
```

Fig. 1.11 Required Python code for generating LIME summary report

```
instance_to_explain = X_train.iloc[23].values

explanation = explainer.explain_instance(
    data_row=instance_to_explain,
    predict_fn=mlp.predict,
    num_features=3
)
print("\n=== LIME Explanation ===")
print("Instance values:", instance_to_explain)
print("Predicted  value:",  mlp.predict(instance_to_explain.reshape(1,  -
1))[0])

print("\nFeature contributions:")
for feature, weight in explanation.as_list():
    print(f"- {feature}: {weight:.4f}")

# 如果在 Jupyter Notebook，可以用視覺化：
# explanation.show_in_notebook(show_table=True, show_all=True)

# import matplotlib.pyplot as plt
# explanation.as_pyplot_figure()
# plt.show()

# pip install html2image
from html2image import Html2Image
explanation.save_to_file("lime_report.html")

hti = Html2Image()

hti.screenshot(
    html_file="lime_report.html",
    save_as="result_summary.png",
    size=(1000, 300)
)
```

Fig. 1.11 (continued)

1.6.2 SHAP Analysis

Post hoc XAI methods apply techniques, such as partial derivation, odd ratio, out-of-bag (OOB) predictor importance, recursive feature elimination (RFE), permutation feature importance (PFI), and SHAP analysis, to evaluate the importance of each input to the output [47], thereby increasing the transparency and interpretability of the ML or DL model, of which SHAP analysis is probably the most popular.

SHAP analysis is a commonly applied XAI technique to evaluate the impact (or importance) of each attribute on the model output. The SHAP value is defined as the weighted average of the marginal contributions of all possible coalitions $|\mathbf{F}|!$ as

$$\varphi_r(f) = \sum_{\{\mathbf{S} \subseteq \mathbf{F}\} \setminus \{r\}} \frac{|\mathbf{S}|!(|\mathbf{F}| - |\mathbf{S}| - 1)!}{|\mathbf{F}|!} \circ \left[f\left(x_{\mathbf{S} \cup \{r\}}\right) - f\left(x_{\mathbf{S}}\right)\right] \tag{1.2}$$

where

$\varphi_r(f)$ is the weighted average Shapley value that feature r provides in the context of all coalitions that exclude feature r;

F is the set of all features;
S is a subset (i.e., coalition) of **F**;
$f\left(x_{\mathbf{S}\cup\{r\}}\right)$ is the model output considering feature r, while $f(x_{\mathbf{S}})$ is the model prediction/output without considering feature r.

A special case is when $|\mathbf{S}| = |\mathbf{F}| - 1$, i.e., $\mathbf{S} = \mathbf{F}\backslash\{r\}$,

$$\varphi_r(f) = \sum_{\{\mathbf{S}\subseteq\mathbf{F}\}\backslash\{r\}} \frac{|\mathbf{S}|!0!}{|\mathbf{F}|!} \circ \left[f\left(x_{\mathbf{S}\cup\{r\}}\right) - f(x_S)\right] = \sum_{\{\mathbf{S}\subseteq\mathbf{F}\}\backslash\{r\}} \frac{\left[f\left(x_{\mathbf{S}\cup\{r\}}\right) - f(x_{\mathbf{S}})\right]}{|\mathbf{F}|!} \quad (1.3)$$

In short, the SHAP value of an attribute refers to the average improvement in performance compared to a fully randomized baseline, given that the value of the attribute is fixed at its current value while other attributes are randomized [15].

In this way, SHAP evaluates the importance of an input by fixing its value while randomizing other inputs [47], and then averaging the resulting differences in $f()$. Take facility location planning problems as an example. Inputs to such an MCDM problem include required investment, probability of success, expected benefits, local acceptance, irreplaceability, etc. [48]. Most of these cannot be meaningfully randomized. The results of a SHAP analysis are often summarized with a force plot or waterfall (or tornado) plot for local explanations or a scatter plot or beeswarm plot for global explanations (see Fig. 1.12). The required Python code is provided in Fig. 1.13.

As can be seen from the figure, the absolute value of the SHAP value of "expected benefits" is the highest, so "expected benefits" is the most important input, followed by "required investment". However, the SHAP value of "required investment" is positive, which means that by setting "required investment" to the current value, the output has increased relative to the average.

```
import xgboost
import shap
import pandas as pd
import numpy as np
VIKOR_results = pd.read_excel('F:\\XAI-class\\講義\\VIKOR-results.xlsx')
X=VIKOR_results.iloc[0:15,0:5]
y=VIKOR_results.iloc[0:15,5:6]
y=y.to_numpy()
model1 = xgboost.XGBRegressor().fit(X, y)
explainer1 = shap.Explainer(model1)
shap_values1 = explainer1(X)
# Explain the first alternative
shap.plots.waterfall(shap_values1[2])
shap.force_plot(shap_values1[2], matplotlib=True)
shap.plots.beeswarm(shap_values1)
shap.plots.scatter(shap_values1[:,          "Expected          benefits"],
color=shap_values1)
```

Fig. 1.12 Required Python code for drawing SHAP plots

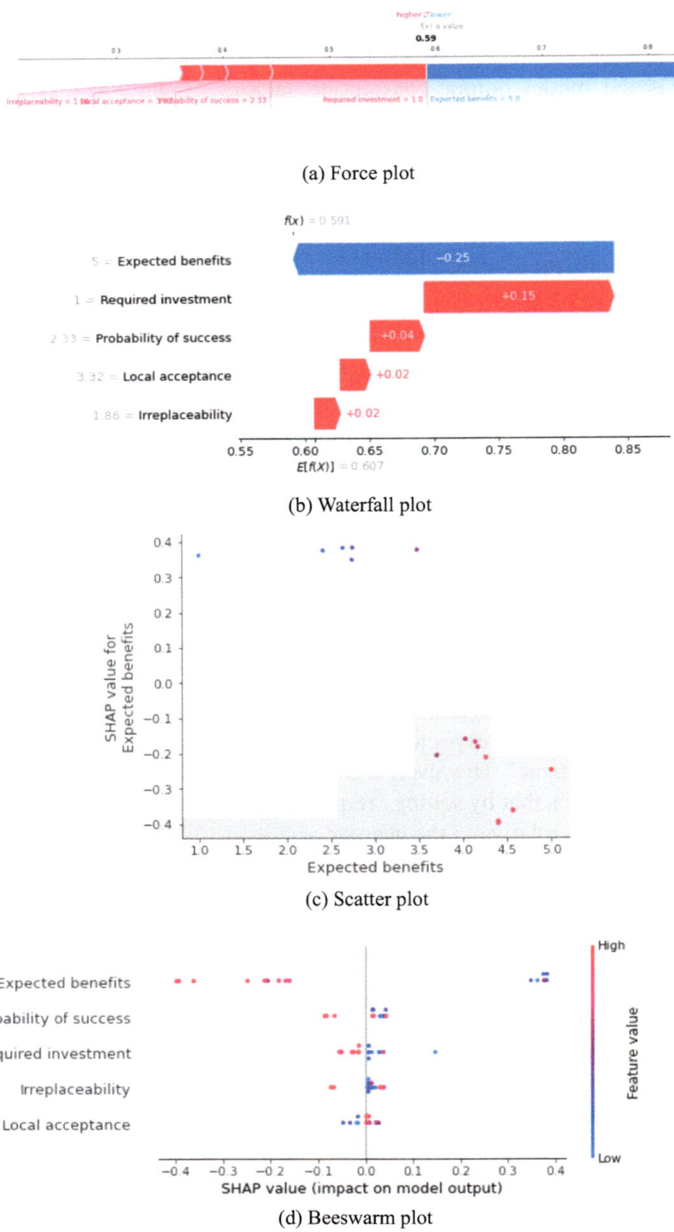

Fig. 1.13 Various SHAP plots

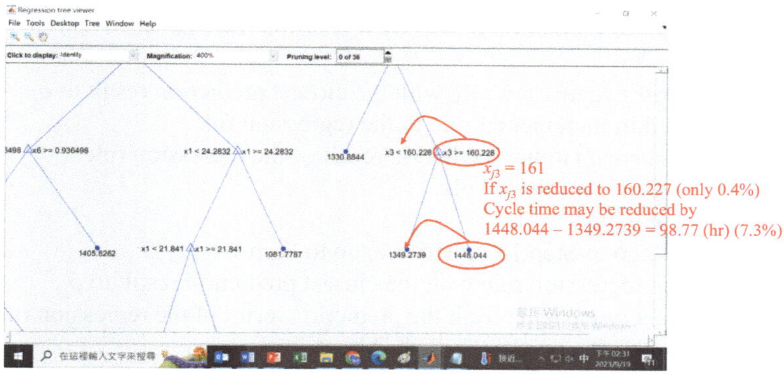

Fig. 1.14 Example of contrastive explanations generated using the local foil tree method

1.6.3 Local Foil Tree Method

The local foil tree method constructs a decision tree centered on any data point in question. Decision trees are trained to locally distinguish foil classes from any other class, including fact classes [49]. The local foil tree method consists of the following steps [49]:

- Retrieve the fact.
- Identify the foil.
- Generate or sample a local data set.
- Train a decision tree.
- Locate the "fact-leaf".
- Locate a "foil-leaf".
- Compute differences.
- Construct explanation.

An example of contrastive explanations generated using the local foil tree method is given in Fig. 1.14.

1.6.4 RF-Based Incremental Interpretation (RFII)

In an RF constructed to approximate a local feasible region, there are a number of trees. From each tree, a regression rule can be applied to predict the cycle time of a job. As a result, multiple rules are simultaneously applicable, which may cause confusion for the user. To address this issue, Chen and Wang [50] proposed RF-based incremental interpretation (RFII) to reorganize the regression rules derived using RF as follows:

Step 1. Average the prediction results by regression rules as $\hat{o}_j$.
Step 2. Set $q = 1$.
Step 3. Choose the regression rule with the closest prediction result to $\hat{o}_j$.
Step 4. Set the q-th incremental rule to the regression rule.
Step 5. Set $\hat{o}_j(current)$ to the prediction result of the regression rule.
Step 6. Remove the regression rule.
Step 7. Increase q by 1.
Step 8. If $q > T$, go to Step 14; otherwise, go to Step 9.
Step 9. Find the regression rule with the closest prediction result to $\hat{o}_j$.
Step 10. Subtract $\hat{o}_j(current)$ from the prediction result of the regression rule and divide the result by T: the result is indicated by $\Delta\hat{o}_j(current)$.
Step 11. Set the premise of the q-th incremental rule to that of the regression rule.
Step 12. Set the consequence of the q-th incremental rule to "add $\Delta\hat{o}_j(current)$ to $\hat{o}_j$" or "subtract $-\Delta\hat{o}_j(current)$ from $\hat{o}_j$".
Step 13. Return to Step 6.
Step 14. End.

1.6.5 Branch-and-Bound (B&B)-Based Decision Rules

A branch-and-bound (B&B) algorithm is traditionally applied to find the optimal solution to an optimization problem, such as those in supply chain planning [51–58], which forms a decision tree through branching and bounding to provide decision rules that can explain the performances of solutions (see Fig. 1.15):

"If $X_{113} = 1$ And $X_{121} = 1$ Then $Z \geq 12(LB)$".

For example, the following B&B-based decision rules are applicable in evaluating supply chain responsiveness:

IF Inventory days of supply is 51.9 AND Delivery cycle time is 4 THEN Responsiveness $\geq$53.47%.

IF Delivery commit date is 98% AND Inventory days of supply is 27.6 AND Delivery cycle time is 1 THEN Responsiveness $\geq$26.84%.

IF Delivery commit date is 85% AND Inventory days of supply is 51.9 AND Delivery cycle time is 1 THEN Responsiveness $\geq$53.47%.

1.7 Organization of This Book

This book is intended to provide technical details on the development and application of XAI to SCM, including methodologies, tools, system architectures, software and hardware, examples and applications. XAI is currently the hottest topic, since the acceptability of overly complex AI technologies is usually questioned. However, most of the AI applications in SCM have not yet been sufficiently explained, which

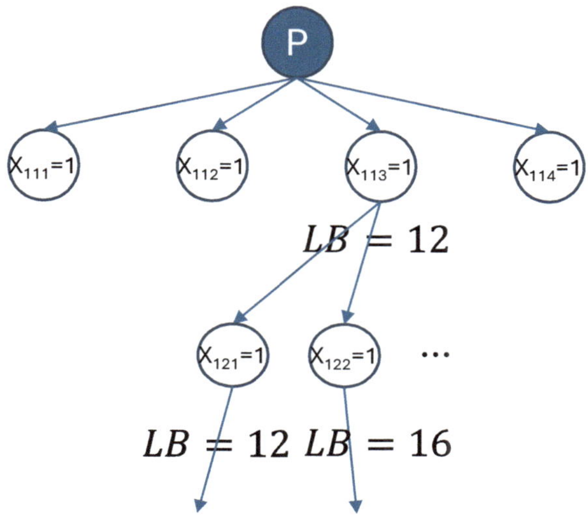

Fig. 1.15 B&B-based decision tree

undoubtedly limits the practicability of these applications. Nevertheless, simple XAI techniques and tools for explaining these AI applications are pursued. However, simple XAI techniques and tools often struggle to adequately explain an AI application. In other words, applying simple XAI techniques and tools to explain a complicated AI application in SCM is still a challenge. However, related research results are mostly scattered in various journal issues or conference proceedings, and there is an urgent need for a systematic integration of these results. To address these issues, this book systematically reviews the progress of XAI and introduces the methods, tools, and applications of XAI technologies in SCM. Relevant references or real cases are also used as supporting evidence.

In specific, the outline of the present book is structured as follows.

The current chapter begins by introducing the basic concepts of XAI and focuses on its application in SCM. First, it defines XAI and highlights some research and development directions in XAI technologies. Then, this chapter elaborates on the implementation process of XAI applications and reviews some important applications in recent literature and reports to illustrate this process. Next, this chapter summarizes the most common XAI applications in SCM, including supply chain product quality management, supply chain resilience, supply chain demand forecasting, supply chain supplier selection, supply chain logistics management, and supply chain performance evaluation. Furthermore, the most commonly used XAI techniques include SHAP analysis and LIME. Next, this chapter discusses the difficulties encountered in applying XAI in SCM. Finally, this chapter introduces some basic XAI techniques and tools, including LIME, SHAP analysis, local foil tree method, RFII, and B&B-based decision rules.

Chapter 2, XAI Applications in Production Planning and Job Scheduling for Supply Chains, begins by defining the fundamental concepts of production planning and job scheduling in supply chains. It then reviews key case studies from relevant literature, summarizing the issues that need to be addressed in supply chain production planning and job scheduling, as well as the methodologies proposed by practitioners and researchers. Through this review, the chapter emphasizes the difficulties in supply chain production planning and scheduling. To overcome these difficulties, various AI technologies have been applied, particularly bio-inspired algorithms such as GAs, ACO, and ABC. The application of these bio-inspired algorithms helps solve optimization problems in production planning or job scheduling for supply chains. However, some AI applications are difficult to understand, communicate, or trust, thus requiring the use of XAI techniques and tools to explain the reasoning processes and results of these AI applications. This chapter introduces the applications of various XAI techniques and tools in production planning and job scheduling for supply chains, such as twin color-coded chromosome maps, contrastive gradient saliency maps, SHAP analysis, and LIME.

Chapter 3, XAI Applications in Facility Location Planning and Supply Network Optimization, first highlights the importance of facility location selection in supply chain management. Subsequently, a hot topic in semiconductor supply chain management, supply chain localization, is introduced, in which semiconductor manufacturing capacity originally owned by wafer foundries is to be transferred back to the home countries of chip makers. Facility location selection becomes especially critical and urgent in semiconductor supply chain localization. AI applications in facility location selection are then reviewed. However, some of these AI applications are not easy to comprehend or communicate. Therefore, XAI techniques and tools are proposed to explain the reasoning process and results of these AI applications, such as hanging gradient bar charts, gradient heatmaps, hanging group gradient bar charts, gradient bidirectional scatterplots, traceable aggregation plot, SHAP analysis, and LIME.

Chapter 4, XAI Applications to Manage Supply Chain Risks and Resilience, first defines the concepts of supply chain risks and supply chain resilience. A trend in the semiconductor industry—semiconductor supply chain localization is owing to pandemic risks, war risks, and political risks. Facing such risks, supply chain participants are taking various actions to mitigate the impact of these risks, so as to maintain their competitiveness and pursue long-term sustainability. To this end, a lot of AI applications can be seen. In particular, there are numerous reports in the literature on the applications of FISs in supply chain risk and resilience management. Therefore, the basic concepts of fuzzy numbers and linguistic terms are introduced. The procedure for building an FIS is also detailed. However, although linguistic terms and FIRs can be easily comprehended, it is still difficult to infer the output of an FIR from the inputs, making the FIS a black box. To address this issue, first, traditional techniques for explaining FISs are reviewed, such as system diagrams and response surface method (RSM). Subsequently, XAI techniques and tools are introduced, including partial dependence plot (PDP), SHAP analysis, and LIME.

References

1. T.C.T. Chen, Y.C. Wang, Artificial intelligence in manufacturing, in *Artificial Intelligence and Lean Manufacturing* (2022), pp. 13–35
2. K.Y. Liu, Introduction, in *Supply Chain Analytics: Concepts, Techniques and Applications* (2022), pp. 1–20
3. H. Min, Artificial intelligence in supply chain management: theory and applications. Int J Log Res Appl **13**(1), 13–39 (2010)
4. A.M. Khedr, Enhancing supply chain management with deep learning and machine learning techniques: a review. J. Open Innovat. Technol. Market Complex. **10**(4), 100379 (2024)
5. M.C. Chiu, T.C.T. Chen, H.C. Wu, Enhancing the understandability and interpretability of type-II diabetes diagnosis using a deep fuzzy neural and XAI approach, in *2025 IEEE 15th Symposium on Computer Applications and Industrial Electronics* (2025), pp. 12–17
6. T.C.T. Chen, M.C. Chiu, Y.C. Lin, Customizable GenAI system for assisting job scheduling in a flexible manufacturing system consisting of multiple factories. Int. J. Adv. Manuf. Technol. **141**, 5721–5734 (2025)
7. A. Daios, N. Kladovasilakis, A. Kelemis, I. Kostavelis, AI applications in supply chain management: a survey. Appl. Sci. **15**(5), 2775 (2025)
8. T.-C. T. Chen, Explainable artificial intelligence (XAI) in manufacturing, in *Explainable Artificial Intelligence (XAI) in Manufacturing: Methodology, Tools, and Applications* (2023), pp. 1–11
9. M. Treppner, H. Binder, M. Hess, Interpretable generative deep learning: an illustration with single cell gene expression data. Hum. Genet. **141**(9), 1481–1498 (2022)
10. D. Gunning, D. Aha, DARPA's explainable artificial intelligence (XAI) program. AI Mag. **40**(2), 44–58 (2019)
11. W. Saeed, C. Omlin, Explainable AI (XAI): a systematic meta-survey of current challenges and future opportunities. Knowl.-Based Syst. **263**, 110273 (2023)
12. M. McNamara, Explainable AI: What is it? How does it work? And what role does data play? (2022). https://www.netapp.com/blog/explainable-ai/
13. C. Panigutti, A. Perotti, D. Pedreschi, Doctor XAI: an ontology-based approach to black-box sequential data classification explanations, in *Proceedings of the 2020 Conference on Fairness, Accountability, and Transparency* (2020), pp. 629–639
14. T.C.T. Chen, C.W. Lin, Y.C. Wang, Visualization XAI techniques for facilitating schedulers' comprehension and interpretation of GA applications in job scheduling. Int. J. Hum.-Comput. Interact. **42**(1), 638–652 (2026)
15. T.C.T. Chen, M.C. Chiu, Y.C. Lin, Novel visualization XAI techniques for assisting schedulers' comprehension and interpretation of ACO applications in job scheduling. Int. J. Hum.–Comput. Interact. 1–34 (2026)
16. M.T. Ribeiro, S. Singh, C. Guestrin, Why should i trust you? Explaining the predictions of any classifier, in *Proceedings of the 22nd ACM SIGKDD International Conference on Knowledge Discovery and Data Mining* (2016), pp. 1135–1144
17. D. Kumar, A. Wong, G.W. Taylor, Explaining the unexplained: A class-enhanced attentive response (clear) approach to understanding deep neural networks, in *Proceedings of the IEEE Conference on Computer Vision and Pattern Recognition Workshops* (2017), pp. 36–44
18. T. Chen, H.-C. Wu, M.-C. Chiu, A deep neural network with modified random forest incremental interpretation approach for diagnosing diabetes in smart healthcare. Appl. Soft Comput. **152**, 111183 (2024)
19. E. Tjoa, H.J. Khok, T. Chouhan, G. Cuntai, Improving deep neural network classification confidence using heatmap-based eXplainable AI (2022). https://doi.org/10.48550/arXiv.2201.000092022
20. J. Lee, H. Cho, Y.J. Pyun, S.J. Kang, H. Nam, Heatmap assisted accuracy score evaluation method for machine-centric explainable deep neural networks. IEEE Access **10**, 64832–64849 (2022)

21. H.-C. Wu, T. Chen, CART–BPN approach for estimating cycle time in wafer fabrication. J. Ambient. Intell. Humaniz. Comput. **6**, 57–67 (2015)
22. Y.-C. Lin, T. Chen, Type-II fuzzy approach with explainable artificial intelligence for nature-based leisure travel destination selection amid the COVID-19 pandemic. Digital Health **8**, 20552076221106320 (2022)
23. J. Senoner, T. Netland, S. Feuerriegel, Using explainable artificial intelligence to improve process quality: evidence from semiconductor manufacturing. Manag. Sci. **68**(8), 5704–5723 (2022)
24. B.O. Kong, M.S. Kim, B.H. Kim, J.H. Lee, Prediction of creep life using an explainable artificial intelligence technique and alloy design based on the genetic algorithm in creep-strength-enhanced ferritic 9% Cr steel. Metals Mater. Int. 1–12 (2023)
25. Y.G. Akhlaghi, K. Aslansefat, X. Zhao, S. Sadati, A. Badiei, X. Xiao, S. Shittu, Y. Fan, X. Ma, Hourly performance forecast of a dew point cooler using explainable artificial intelligence and evolutionary optimisations by 2050. Appl. Energy **281**, 116062 (2021)
26. T.C.T. Chen, Qualitative and quantitative analysis of semiconductor supply chain localization, in *Supply Chain Localization in the Semiconductor Industry: Rebuilding the Competitiveness and Sustainability of Semiconductor Manufacturers* (2025), pp. 73–125
27. T.C.T. Chen, Explaining genetic algorithm applications in job sequencing and scheduling, in *Explainable and Customizable Job Sequencing and Scheduling: Advancing Production Control and Management with XAI* (2025), pp. 73–102
28. B. Mahbooba, M. Timilsina, R. Sahal, M. Serrano, Explainable artificial intelligence (XAI) to enhance trust management in intrusion detection systems using decision tree model. Complexity **2021**(1), 6634811 (2021)
29. G. Mugurusi, P.N. Oluka, Towards explainable artificial intelligence (xai) in supply chain management: A typology and research agenda, in *IFIP International Conference on Advances in Production Management Systems* (2021), pp. 32–38
30. S. Bhatia, A.S. Albarrak, A blockchain-driven food supply chain management using QR code and XAI-faster RCNN architecture. Sustainability **15**(3), 2579 (2023)
31. S. Schoepf, J. Foster, A. Brintrup, Identifying contributors to supply chain outcomes in a multiechelon setting: a decentralised approach. IEEE Trans. Indus. Inform.(2024)
32. K. Sadeghi, D. Ojha, P. Kaur, R.V. Mahto, A. Dhir, Explainable artificial intelligence and agile decision-making in supply chain cyber resilience. Decis. Support. Syst. **180**, 114194 (2024)
33. S.K. Jauhar, S. Harinath, V. Krishnaswamy, S.K. Paul, Explainable artificial intelligence to improve the resilience of perishable product supply chains by leveraging customer characteristics. Ann. Oper. Res. **354**(1), 103–142 (2025)
34. S. Alhazmi, M.K.I. Rahmani, M. Arif, M.T. Nafis, Developing intelligent and immutable vaccine supply and operation platform using blockchain and artificial intelligence technologies. IEEE Access **12**, 88189–88201 (2024)
35. T.C.T. Chen, Y.C. Wang, Y.C. Wang, An explainable decision model for selecting facility locations in supply chain networks. Supply Chain Anal. **11**, 100148 (2025)
36. Y.C. Wang, T.C.T. Chen, C.W. Lin, Fuzzy reinforcement learning approach for supporting ubiquitous manufacturing facility selection in drone manufacturing. Int. J. Adv. Manuf. Technol. **141**(3), 1893–1903 (2025)
37. Y.C. Wang, T. Chen, New XAI tools for selecting suitable 3D printing facilities in ubiquitous manufacturing. Complex Intell. Syst. **9**(6), 6813–6829 (2023)
38. U. Kamath, J. Liu, *Explainable Artificial Intelligence: An Introduction to Interpretable Machine Learning* (Springer, 2021)
39. T.C.T. Chen, Management support, in *Advances in Cycle Time Management* (2025), pp. 97–135
40. Y.C. Wang, T. Chen, M.C. Chiu, An explainable deep-learning approach for job cycle time prediction. Decis. Anal. J. **6**, 100153 (2023)
41. K.R. Ahmed, M.E. Ansari, M.N. Ahsan, A. Rohan, M.B. Uddin, M.A.H. Rivin, Deep learning framework for interpretable supply chain forecasting using SOM ANN and SHAP: KR Ahmed et al. Sci Rep 15(1):26355 (2025)

42. S. Jaipuria, S.S. Mahapatra, An improved demand forecasting method to reduce bullwhip effect in supply chains. Expert Syst. Appl. **41**(5), 2395–2408 (2014)
43. R.F. Saen, F. Yousefi, M. Azadi, Artificial intelligence powered predictions: enhancing supply chain sustainability. Ann Oper Res 1–44 (2024)
44. A.M. Salih, Z. Raisi-Estabragh, I.B. Galazzo, P. Radeva, S.E. Petersen, K. Lekadir, G. Menegaz, A perspective on explainable artificial intelligence methods: SHAP and LIME. Adv. Intell. Syst. **7**(1), 2400304 (2025)
45. E. Pourjavad, A. Shahin, The application of Mamdani fuzzy inference system in evaluating green supply chain management performance. Int. J. Fuzzy Syst. **20**(3), 901–912 (2018)
46. G.M.D. Ganga, L.C.R. Carpinetti, A fuzzy logic approach to supply chain performance management. Int. J. Prod. Econ. **134**(1), 177–187 (2011)
47. T.-C.T. Chen, Applications of XAI for forecasting in the manufacturing domain, in *Explainable Artificial Intelligence (XAI) in Manufacturing: Methodology, Tools, and Applications* (2023), pp. 13–50
48. T.C.T. Chen, *Supply Chain Localization in the Semiconductor Industry* (Springer Nature, 2025)
49. J. van der Waa, M. Robeer, van J. Diggelen, M. Brinkhuis, M. Neerincx, Contrastive explanations with local foil trees (2018). arXiv:1806.07470
50. T. Chen, Y.C. Wang, A two-stage explainable artificial intelligence approach for classification-based job cycle time prediction. Int. J. Adv. Manuf. Technol. **123**(5), 2031–2042 (2022)
51. S.H. Zegordi, I.K. Abadi, M.B. Nia, A novel genetic algorithm for solving production and transportation scheduling in a two-stage supply chain. Comput. Ind. Eng. **58**(3), 373–381 (2010)
52. D. Ivanov, B. Sokolov, Dynamic supply chain scheduling. J. Sched. **15**(2), 201–216 (2012)
53. M. Musavi, A. Bozorgi-Amiri, A multi-objective sustainable hub location-scheduling problem for perishable food supply chain. Comput. Ind. Eng. **113**, 766–778 (2017)
54. G. Wang, Integrated supply chain scheduling of procurement, production, and distribution under spillover effects. Comput. Oper. Res. **126**, 105105 (2021)
55. C. Pérez, L. Climent, G. Nicoló, A. Arbelaez, M.A. Salido, A hybrid metaheuristic with learning for a real supply chain scheduling problem. Eng. Appl. Artif. Intell. **126**, 107188 (2023)
56. A. Borumand, M.A. Beheshtinia, A developed genetic algorithm for solving the multi-objective supply chain scheduling problem. Kybernetes **47**(7), 1401–1419 (2018)
57. F. Xiong, P. Gong, P. Jin, J.F. Fan, Supply chain scheduling optimization based on genetic particle swarm optimization algorithm. Clust. Comput. **22**(Suppl 6), 14767–14775 (2019)
58. I. Dzalbs, T. Kalganova, Accelerating supply chains with ant colony optimization across a range of hardware solutions. Comput. Ind. Eng. **147**, 106610 (2020)

Chapter 2
XAI Applications in Production Planning and Job Scheduling for Supply Chains

2.1 Production Planning and Job Scheduling

There are three flows in a supply chain: material flow, information flow, and finance flow [1]. Production planning and job scheduling are critical tasks to maintaining smooth material flow. Production planning in a supply chain is to determine multi-period and multi-stage master production plans for manufacturers, along with resource levels and resource allocations, to minimize manufacturing costs [2]. Job scheduling refers to allocating a given set of jobs to a given set of resources while meeting specific objectives [3]. Job scheduling is one of the most popular and fundamentally sound research fields in operations research, production, logistics, and supply chain management (SCM). Compared to production planning, job scheduling typically focuses on the short term. Furthermore, production planning requires forecasting product demand, while job scheduling is usually based on already forecasted product demand and only requires forecasting some short-term operatable time measures, such as the remaining cycle time of a job [4]. Nevertheless, both production planning and job scheduling involve the formulation and solving of optimization problems.

2.2 Production Planning and Job Scheduling in Supply Chains

The following reviews some literature on production planning and job scheduling in supply chains.

Zegordi et al. [5] addressed the scheduling problem for a two-echelon supply chain consisting of manufacturers and transport vehicles. The manufacturers had varying production rates, and the transport vehicles had unequal speeds and capacities. This supply chain scheduling problem was formulated as a mixed-integer linear

T.-C. T. Chen, *Explainable Artificial Intelligence in Supply Chain Management*,
SpringerBriefs in Applied Sciences and Technology,
https://doi.org/10.1007/978-3-032-22564-1_2

programming (MILP) problem and solved using a gendered genetic algorithm (GA) involving two different chromosomes with unequal structures.

Jauhar and Pant [6] applied a software package developed in Excel to plan customer orders in a supply chain, where a multi-objective optimization problem was solved to maximize the service level, minimize penalties, and minimize inequality in resource use by using the optimal program control method in which the Hamiltonian was maximized at each instant of time.

Jauhar and Pant [6] categorized the applications of GAs in supply chain management into eight types: production process management, order fulfillment, demand management, supplier relationship management, product development and commercialization, returns management, customer service management, and customer relationship management. Among these, job scheduling using GAs was particularly critical for production process management and order fulfillment.

Musavi et al. [7] considered a perishable food supply chain with a limited number of transport fleets serving all hubs, necessitating allocation and scheduling. A multi-objective MILP model was formulated to minimize total transportation costs, maximize the freshness and quality of food at the time of delivery, and minimize the total carbon emissions of the vehicles. The non-dominated sorting genetic algorithm-II (NSGA-II) was applied to solve the multi-objective scheduling problem.

Wang [8] addressed the job scheduling problem of a three-echelon supply chain involving raw material suppliers, manufacturers, and retailers. The scheduling objective was to minimize total cost, including transportation costs and penalties that were nonlinearly related to the retailer's unfilled quantity and fill rate. A mixed-integer nonlinear programming (MINLP) model was formulated and solved by incorporating the augmented Lagrangian relaxation method and a particle swarm optimization (PSO) algorithm.

Pérez et al. [9] solved a supply chain scheduling problem that assigned orders to available warehouses and completed their loading within a week. The goal was to minimize total transportation cost and the cost of producing additional inventory to meet weekly demand. The optimization problem was solved using a hybrid approach that incorporated two metaheuristic algorithms: the greedy randomized adaptive search program (GRASP) and a GA. To this end, various software packages were applied and compared: GECODE 6.3.0, YUCK 2023, OR-TOOLS 9.6, and IBM ILOG CP optimizer.

Borumand and Beheshtinia [10] constructed a multi-objective MILP model for production planning and transportation scheduling in a two-tier supply chain involving a manufacturer and its suppliers. The model aimed to minimize total production costs, delivery tardiness, and emissions while maximizing product quality. Instead of seeking Pareto optimal solutions to the multi-objective MILP problem, they used a well-known multi-criteria decision-making method—višekriterijumska optimizacija i kompromisno rešenje (VIKOR)—to evaluate the overall performance of each feasible solution. Then, using the overall performance of each feasible solution as its fitness, a GA was proposed to evolve a set of feasible solutions, ultimately yielding the optimal solution.

Xiong et al. [11] solved the production planning problem in a two-level supply chain, aiming to minimize total production costs. To this end, they employed a hybrid approach combining GA and particle swarm optimization (PSO). In the hybrid approach, each population was evolved using both of these bio-inspired algorithms, and the results were then optimally combined to form the next population.

Transportation planning is a critical task in logistics supply chains and is often modeled as a traveling salesman problem (TSP). To improve the efficiency of solving such problems in practical applications, Dzalbs and Kalganova [12] proposed a parallel ACO method. Compared to independent ACO, parallel ACO was able to achieve higher-quality solutions in fewer iterations as the number of parallel ant colonies increased. Furthermore, they tested the speed performance under three different hardware configurations. The results showed that GPUs were not suitable for transportation planning in complex real-world supply chains due to the need to access large amounts of metadata.

Ubiquitous manufacturing (UM) systems are typical two-echelon or three-echelon supply chains involving a number of manufacturers. Wang et al. [13] established a UM system of 3D printing facilities to meet the explosive growth demand for drones, in which a fuzzy reinforcement learning approach was proposed to plan the operations in the UM system. The fuzzy reinforcement learning approach employed the Q-learning algorithm to accelerate the prioritizing of criteria for 3D printing facility selection. Subsequently, fuzzy technique for order preference by similarity to the ideal solution (FTOPSIS) was applied to compare various 3D printing facilities to ensure that drone orders were correctly assigned to suitable UM partners. In contrast to the GA employed in previous research, the fuzzy reinforcement learning method was model-free and particularly efficient in unpredictable or random settings.

2.3 Difficulties of Production Planning and Job Scheduling in Supply Chains

Existing methods in this field face several challenges (see Fig. 2.1). The first challenge was constructing the required optimization models. Most of these optimization problems were also difficult to solve using existing optimization software packages (i.e., NP-hard), leading to the second challenge. Researchers needed to find new ways to apply existing algorithms or devise new algorithms to solve this problem. This is the third challenge. The entire process requires a high degree of professional knowledge and even expertise, far beyond the capabilities of factory schedulers. This is the fourth challenge.

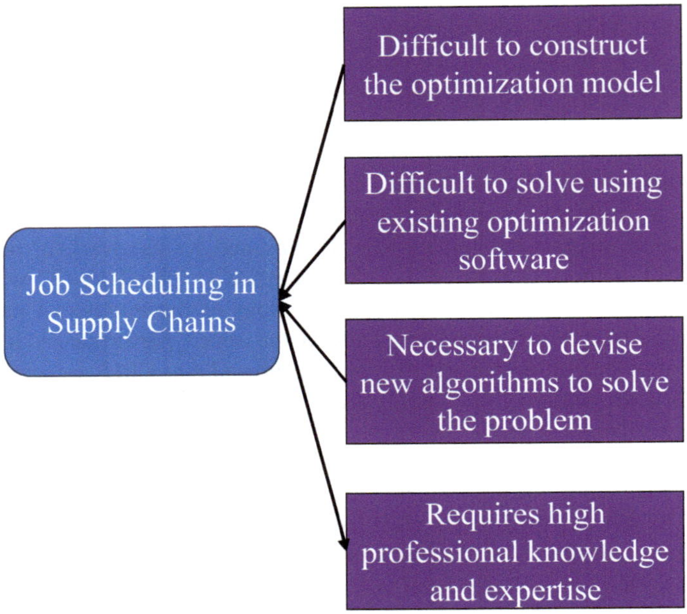

Fig. 2.1 Difficulties in production planning and job scheduling in supply chains

2.4 AI Applications for Production Planning in Supply Chains

Chen [14] made the production plan of a UM system, which was a special three-tier supply chain (see Fig. 2.2). In the production planning method, two optimization problems were formulated: fuzzy mixed-integer linear programming (FMILP) and fuzzy mixed-integer quadratic programming (FMIQP) problems, highlighting the prevalence of applying fuzzy logic in tackling uncertainty in supply chain planning.

Subsequently, slack was derived to determine whether to restart an early terminated 3D printing process in the same 3D printing facility, enhancing the supply chain resilience further. The objective function of the FMILP model is to minimize the fuzzy makespan:

(FMILP Model)

$$\text{Min } \tilde{Z}_1 = \max_i(\tilde{a}_i(+)n_i\tilde{p}_i) \tag{2.1}$$

where

$$\sum_{n}^{i=1} n_i = N \tag{2.2}$$

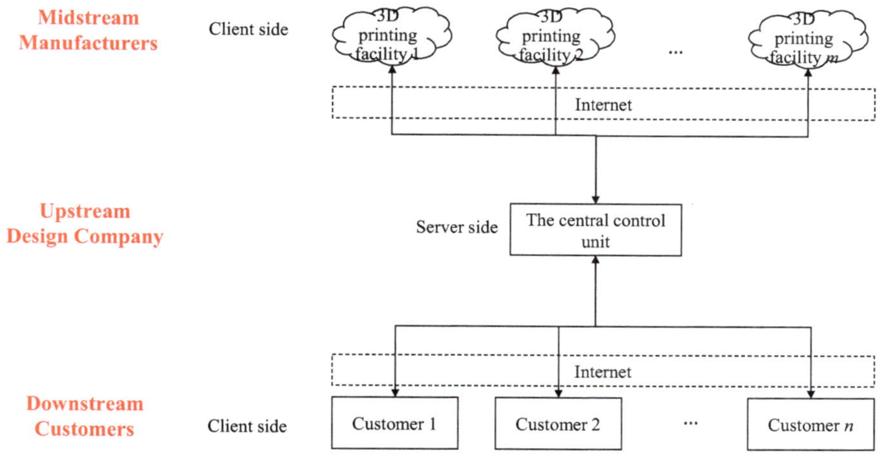

Fig. 2.2 UM system as a special three-tier supply chain

$$n_i \in \mathbf{Z}^+ \cup \{0\};\ i = 1 \sim m \tag{2.3}$$

Here, $\tilde{a}_i$ is the available time of the ith 3D printing facility, $i = 1 \sim m$; $\tilde{p}_i$ is the time required to print a piece in the ith 3D printing facility; $\tilde{a}_i(+)n_i\tilde{p}_i$ is the (fuzzy) completion time of the ith 3D printing facility that prints n_i pieces; and N is the total number of pieces.

The objective function $\tilde{Z}_1 = (Z_{11}, Z_{12}, Z_{13})$ is to minimize the fuzzy maximal completion time, that is, the fuzzy makespan. In the literature, a variety of methods for dealing with a fuzzy objective function have been presented. For example, a fuzzy objective function can be replaced with three crisp objective functions: minimizing the core (i.e., Z_{12}), maximizing the left tail (i.e., $Z_{12} - Z_{11}$), and minimizing the right tail (i.e., $Z_{13} - Z_{12}$). The desirable region for each objective function is specified in the form of a trapezoidal fuzzy number to evaluate the satisfaction level to be maximized. First, $1 \leq n_i \leq N$ and min() $\leq$ max(). Therefore,

$$\min_i(a_{i2} + p_{i2}) \leq Z_{12} = \max_i(a_{i2} + n_i p_{i2}) \leq \max_i(a_{i2} + Np_{i2}) \tag{2.4}$$

as illustrated in Fig. 2.3.

Similarly,

$$\min_i(a_{i2} + p_{i2} - a_{i1} - p_{i1}) \leq Z_{12} - Z_{11} = \max_i(a_{i2} + n_i p_{i2}) - \max_i(a_{i1} + n_i p_{i1}) \leq \max_i(a_{i2} + Np_{i2} - a_{i1} - Np_{i1}) \tag{2.5}$$

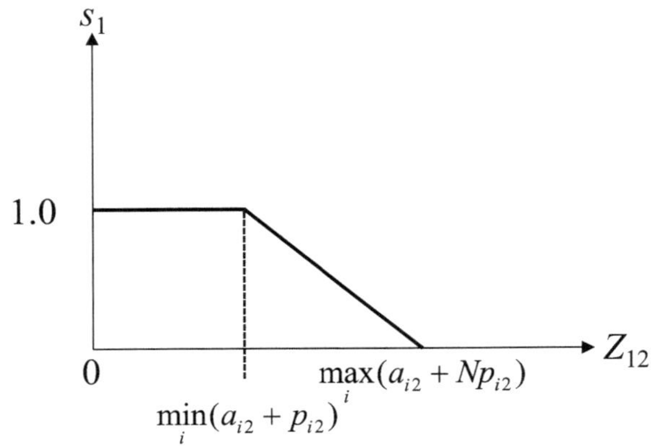

Fig. 2.3 Trapezoidal satisfaction region for Z_{12}

as illustrated in Fig. 2.4.

In addition,

$$\min_i(a_{i3} + p_{i3} - a_{i2} - p_{i2}) \leq Z_{13} - Z_{12} = \max_i(a_{i3} + n_i p_{i3}) - \max_i(a_{i2} + n_i p_{i2}) \leq \max_i(a_{i3} + Np_{i3} - a_{i2} - Np_{i2}) \quad (2.6)$$

as illustrated in Fig. 2.5.

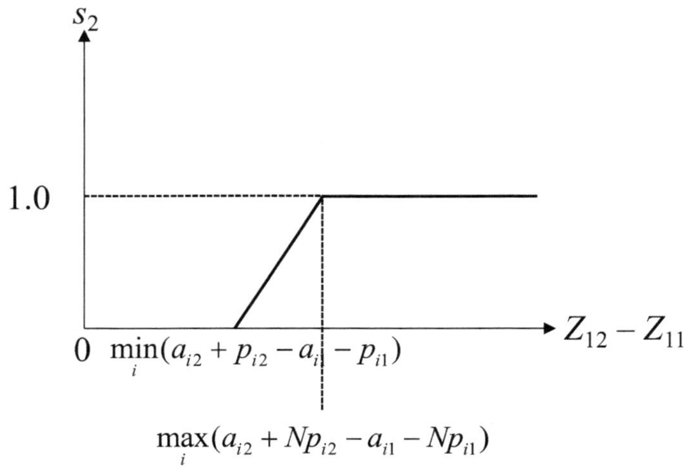

Fig. 2.4 Trapezoidal satisfaction region for $Z_{12} - Z_{11}$

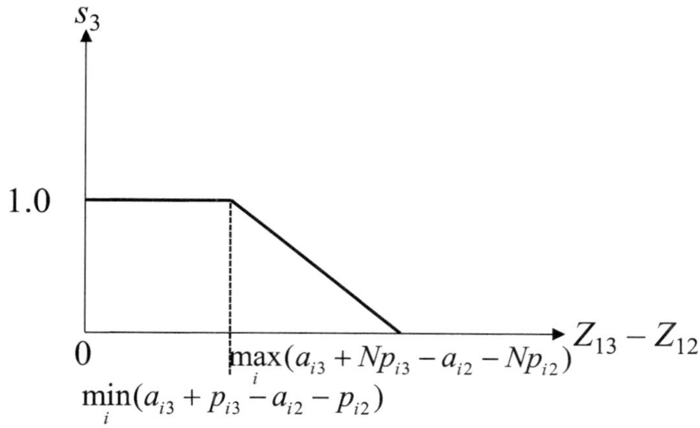

Fig. 2.5 Trapezoidal satisfaction region for $Z_{13} - Z_{12}$

The objective function (2.1) is converted into the following:

$$\text{Max } s_1 \tag{2.7}$$

where

$$s_1 \leq s_{11} \tag{2.8}$$

$$s_1 \leq s_{12} \tag{2.9}$$

$$s_1 \leq s_{13} \tag{2.10}$$

$$s_{11} \leq \frac{\max_i(a_{i2} + N \cdot p_{i2}) - Z_{12}}{\max_i(a_{i2} + N \cdot p_{i2}) - \min_i(a_{i2} + p_{i2})} \tag{2.11}$$

$$s_{12} \leq \frac{Z_{12} - Z_{11} - \min_i(a_{i2} + p_{i2} - a_{i1} - p_{i1})}{\max_i(a_{i2} + N \cdot p_{i2} - a_{i1} - N \cdot p_{i1}) - \min_i(a_{i2} + p_{i2} - a_{i1} - p_{i1})} \tag{2.12}$$

$$s_{13} \leq \frac{\max_i(a_{i3} + N \cdot p_{i3} - a_{i2} - Np_{i2}) - (Z_{13} - Z_{12})}{\max_i(a_{i3} + N \cdot p_{i3} - a_{i2} - N \cdot p_{i2}) - \min_i(a_{i3} + p_{i3} - a_{i2} - p_{i2})} \tag{2.13}$$

where s_1 is the degree to which the goal of the fuzzy objective function ($\tilde{Z}_1$) is satisfied. Moreover, s_{11}, s_{12}, and s_{13} indicate the degrees to which Z_{12} is minimized, $Z_{12} - Z_{11}$ is maximized, and $Z_{13} - Z_{12}$ is minimized, respectively.

Subsequently, the three satisfaction levels can be easily aggregated. However, several subproblems of the FMILP problem must be solved, and thus this method may not be efficient for online applications.

The optimal solution of the FMILP model also provides slack information to determine whether an early terminated 3D printing process should be restarted. Based on the optimal production plan generated by the FMILP model, the FMIQP model prepares a transportation plan to collect the printed pieces as soon as possible. The FMILP and FMIQP models are converted into equivalent mixed-integer linear programming (MILP) and mixed-integer quadratic programming (MIQP) problems, respectively, to be more easily solved. Both MILP and MIQP problems are NP-hard [15, 16]. In particular, the MIQP problem is difficult to solve.

(MIQP Model)

$$\text{Max } s_2 \tag{2.14}$$

where

$$s_2 \leq s_{21} \tag{2.15}$$

$$s_2 \leq s_{22} \tag{2.16}$$

$$s_2 \leq s_{23} \tag{2.17}$$

$$s_{21} \leq \frac{Z_{12}^* + (n-1)\max_{i \neq j} d_{ij} + 2\max_i d_{Oi} - Z_{22}}{(n-1)\max_{i \neq j} d_{ij} + 2\max_i d_{Oi}} \tag{2.18}$$

$$s_{22} \leq \frac{Z_{22} - Z_{21} - Z_{12}^* + Z_{11}^* + (n-1)\max_{i \neq j} d_{ij} + 2\max_i d_{Oi}}{2(n-1)\max_{i \neq j} d_{ij} + 4\max_i d_{Oi}} \tag{2.19}$$

$$s_{23} \leq \frac{Z_{13}^* + (n-1)\max_{i \neq j} d_{ij} + 2\max_i d_{Oi} - Z_{12}^* - Z_{23} + Z_{22}}{2(n-1)\max_{i \neq j} d_{ij} + 4\max_i d_{Oi}} \tag{2.20}$$

$$Z_{2k} = \sum_{i=1}^{m} X_{iO}(l_{ik} + d_{iO});\ k = 1 \sim 3 \tag{2.21}$$

$$r_{i1} \geq X_{Oi}(t + d_{Oi});\ i = 1 \sim m \tag{2.22}$$

$$r_{ik} \geq X_{ji}(l_{jk} + d_{ji});\ i, j = 1 \sim m;\ j \neq i;\ k = 1 \sim 3 \tag{2.23}$$

$$l_{ik} \geq a_{ik} + n_i^* p_{ik};\ i = 1 \sim m;\ k = 1 \sim 3 \tag{2.24}$$

$$l_{ik} \geq r_{ik};\ i = 1 \sim m;\ k = 1 \sim 3 \tag{2.25}$$

$$X_{ij} + X_{ji} \leq 1;\ i, j = 1 \sim m;\ j \neq i \tag{2.26}$$

$$X_{Oi} + \sum_{j \neq i} X_{ji} = 1;\ i = 1 \sim m \tag{2.27}$$

$$X_{iO} + \sum_{j \neq i} X_{ij} = 1;\ i = 1 \sim m \tag{2.28}$$

$$\sum_{i=1}^{m} X_{Oi} = 1 \tag{2.29}$$

$$\sum_{i=1}^{m} X_{iO} = 1 \tag{2.30}$$

$$0 \leq Z_{21} \leq Z_{22} \leq Z_{23} \tag{2.31}$$

$$0 \leq r_{i1} \leq r_{i2} \leq r_{i3};\ i = 1 \sim m \tag{2.32}$$

$$0 \leq l_{i1} \leq l_{i2} \leq l_{i3};\ i = 1 \sim m \tag{2.33}$$

$$X_{Oi},\ X_{iO},\ X_{ij} \in \{0,\ 1\};\ i, j = 1 \sim m;\ j \neq i \tag{2.34}$$

where the objective function is to maximize the minimum satisfaction level of three goals in optimizing two fuzzy objectives: $\tilde{Z}_1$ is to minimize the fuzzy maximal completion time, that is, the fuzzy makespan; $\tilde{Z}_2$ is to minimize the fuzzy cycle time determined using the leaving time of the freight vehicle from the most recently visited 3D printing facility; n is the number of 3D printing facilities; n_i is the number of pieces assigned to 3D printing facility i; d_{ij} is the shortest path length between 3D printing facilities i and j, $j = 1 \sim m$, $j \neq I$, $d_{ij} = d_{ji}$; d_{Oi} is the shortest path length between O and 3D printing facility i; $\widetilde{r}_i$ is the arrival time at the ith 3D printing facility; $\widetilde{l}_i$ is the time at which the freight truck leaves the ith 3D printing facility; and X_{ij} is a state variable. If the freight truck travels from 3D printing facility i to 3D printing facility j, $X_{ij} = 1$; otherwise, $X_{ij} = 0$, where $j = 1 \sim m$ and $j \neq i$. X_{iO} is a state variable. If the freight truck returns to O from 3D printing facility i, then $X_{iO} = 1$; otherwise, $X_{iO} = 0$. X_{Oi} is a state variable. If the freight truck travels from O to 3D printing facility i, then $X_{Oi} = 1$; otherwise, $X_{Oi} = 0$.

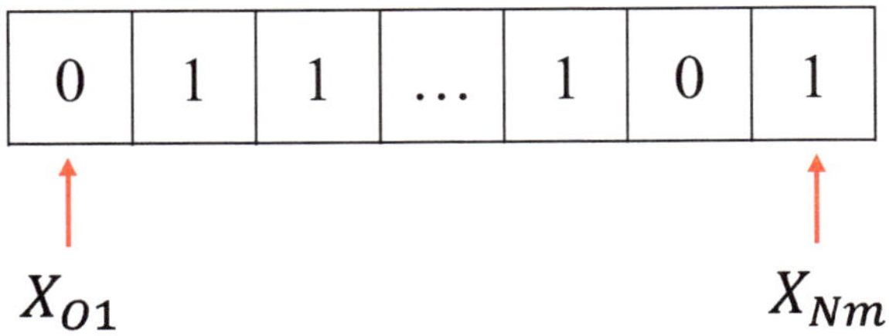

Fig. 2.6 Chromosome structure in the GA for solving the MIQP problem

GA can be applied to help solve the MIQP problem. Basically, the values of the state variables are to be determined. Therefore, chromosomes in the GA may comprise binary arrays of the state variables, as illustrated in Fig. 2.6.

2.5 AI Applications for Job Scheduling in Supply Chains

2.5.1 Production Planning Model of a Three-Echelon Supply Chain

Take a three-echelon supply chain as an example, which comprises raw material suppliers, factories, and customers (or retailers), as illustrated in Fig. 2.7. Factories in the three-echelon supply chain are flexible job shops to provide great manufacturing flexibility.

The production and transportation planning problem of a three-echelon supply chain can be formulated as an optimization problem, in which the following variables are used:

- α_{js}: $\alpha_{js} = 1$ if the raw materials for making job j are supplied by supplier s; otherwise, $\alpha_{js} = 0.j = 1 \sim n$, $s = 1 \sim S$.
- β_{jf}: $\beta_{jf} = 1$ if job j is assigned to factory f; otherwise, $\beta_{jf} = 0.j = 1 \sim n, f = 1 \sim F$.
- γ_{jc}: $\gamma_{jf} = 1$ if job j belongs to customer c; otherwise, $\gamma_{jc} = 0.j = 1 \sim n$, $c = 1 \sim C$.
- ψ_{sf}: the transportation cost between supplier s and factory $f.s = 1 \sim S, f = 1 \sim F$.
- ψ_{fc}: the transportation cost between factory f and customer $c. f = 1 \sim F, c = 1 \sim C$
- TP_j: the transportation cost of job (order) $j.j = 1 \sim n$.
- L_j: the lateness of job j.
- C_j: the completion time of job j.
- d_j: the due date of job j.
- T_j: the tardiness of job j.
- ρ_j: the unit penalty for job j in case of late delivery. $j = 1 \sim n$.
- PE_j: the penalty cost of job (order) $j.j = 1 \sim n$.

- Φ_s: the number of jobs supported by supplier s. $s = 1 \sim S$.
- s_{jk}: the start time of operation k of job j.
- c_{jk}: the completion time of operation k of job j.
- ω_1, ω_2: coefficients.
- X_{jkfc}: $X_{jkfc} = 1$ if job j belongs to customer c; otherwise, $X_{jkfc} = 0.j = 1 \sim n, k = 1 \sim K, f = 1 \sim F, c = 1 \sim C$.
- $Y_{j_1k_1j_2k_2}$: $Y_{j_1k_1j_2k_2} = 1$ if operation k_1 of job j_1 precedes operation k_2 of job j_2 on the same machine; otherwise, $Y_{j_2k_2j_1k_1} = 1$. $1 \le j_1, j_2 \le n, j_1 \ne j_2, 1 \le k_1, k_2 \le K$.

In addition, two assumptions are made:

- It is assumed that all operations of a job are performed in the same factory.
- Machines in all factories are available at the start of the planning horizon, i.e., $a_m = 0\ \forall m$.

The production and transportation planning model aims to minimize the total cost (TC) of fulfilling n orders through the three-echelon flexible supply chain:

$$\text{Min } TC = \omega_1 \sum_{j=1}^{n} TP_j + \omega_2 \sum_{j=1}^{n} PE_j \tag{2.35}$$

s.t.

$$\sum_{s=1}^{S} \alpha_{js} = 1; j = 1 \sim n \tag{2.36}$$

$$\sum_{f=1}^{F} \beta_{jf} = 1; j = 1 \sim n \tag{2.37}$$

$$TP_j = \sum_{f=1}^{F} \sum_{s=1}^{S} \alpha_{js}\beta_{jf}\psi_{sf} + \sum_{f=1}^{F} \beta_{jf}\gamma_{jc}\psi_{fc}; j = 1 \sim n \tag{2.38}$$

$$L_j = C_j - d_j; j = 1 \sim n \tag{2.39}$$

$$T_j = \max\left(L_j, 0\right); j = 1 \sim n \tag{2.40}$$

$$PE_j = \rho_j T_j; j = 1 \sim n \tag{2.41}$$

$$\Phi_s = \sum_{j=1}^{n} \alpha_{js}; s = 1 \sim S \tag{2.42}$$

$$\Phi_s \le \pi_s; s = 1 \sim S; \tag{2.43}$$

$$\sum_{s=1}^{S} \Phi_s = n \tag{2.44}$$

$$s_{j1} \geq \sum_{f=1}^{F} \sum_{s=1}^{S} \alpha_{js} \beta_{jf} \psi_{sf} + r_j; j = 1 \sim ns \tag{2.45}$$

$$c_{jk} = s_{jk} + \sum_{f=1}^{F} \sum_{i=1}^{m} \beta_{jf} X_{jkfi} p_{jkfi}; j = 1 \sim n, k = 1 \sim K \tag{2.46}$$

$$C_j = c_{jK}; j = 1 \sim n \tag{2.47}$$

$$\sum_{f=1}^{F} \sum_{i=1}^{m} \beta_{jf} X_{jkfi} = 1; j = 1 \sim n, k = 1 \sim K \tag{2.48}$$

$$\begin{aligned} s_{j_2k_2} \geq \sum_{f=1}^{F} \sum_{i=1}^{m} \beta_{j_1f} X_{j_1k_1fi} \beta_{j_2f} X_{j_2k_2fi} Y_{j_1k_1j_2k_2} c_{j_1k_1} \\ ; \forall 1 \leq k_1, k_2 \leq K, 1 \leq j_1, j_2 \leq n, j_1 \neq j_2 \end{aligned} \tag{2.49}$$

$$Y_{j_1k_1j_2k_2} + Y_{j_2k_2j_1k_1} = 1; \forall 1 \leq k_1, k_2 \leq K, 1 \leq j_1, j_2 \leq n, j_1 \neq j_2 \tag{2.50}$$

$$\begin{aligned} X_{jkfi}, Y_{j_1k_1j_2k_2} \in \{0, 1\}; \forall 1 \leq k, k_1, k_2 \leq K, 1 \leq j, j_1, j_2 \leq n, \\ j_1 \neq j_2, i = 1 \sim m, f = 1 \sim \eta_f \end{aligned} \tag{2.51}$$

Other variables $\in R^+$

According to Eq. (2.36), the raw materials for each job (order) can only be provided by a single supplier. In addition, each job can only be assigned to a single factory, as requested by Eq. (2.35). The transportation cost of a job is calculated according to Eq. (2.38). The penalty cost of each job (order) is derived by Eq. (2.28), for which the tardiness is evaluated using Eqs. (2.39)–(2.40). Equation (2.42) summarizes the burden on a supplier, which is bounded, as shown in Eq. (2.43). Equation (2.44) depicts that all suppliers collaborate to support these jobs. Once job j is released (i.e., the order has been placed) and the required raw materials have been delivered to the selected factory, its first operation can begin, as described in Eq. (2.45). The completion time of an operation depends on the factory and the machine on which the operation is performed, as derived by Eq. (2.46). According to Eq. (2.47), the completion time of a job depends on its last operation. Constraint (2.48) requires that an operation can only be performed on a single machine in a factory. Constraint (2.49) sequences two operations if they are performed on the same machine of a factory.

Chen [17] proposed a five-field notation (FFN) to describe the job scheduling problem in a four-echelon supply chain as manufacturing system type/production conditions/delivery/number of customers/scheduling objective. In this study, for a three-echelon supply chain without delivery service providers, the production and transportation planning model can be indicated with the five-field notation: $FJm/r_j//\eta_c/TC$.

The production and transportation planning model of a three-echelon supply chain is an MINLP problem, which is NP-hard, even harder than the scheduling problem of a single flexible job shop [18] or a flexible manufacturing system [19]. In addition, the default objective function is to minimize total cost, while in flexible job shops or flexible manufacturing systems the default objective function is usually the minimization of the makespan [20].

2.5.2 AI Application

Deriving the global optimal solution to the MINLP problem is difficult, a GA can be used to help solve the MINLP problem.

There are two sets of chromosomes in each population. The first set determines the supplier of each job; the second set assigns operations to machines. Each chromosome (i.e., job permutation or feasible solution) is represented as an array string called chromosome diagrams (or maps), as illustrated in Fig. 2.8. In this figure, the array (j, o, m) means that the o-th operation of job j was performed on machine m.

The procedure for solving a scheduling problem using GA is composed of the following major steps (see Fig. 2.9):

- First, the chromosomes of the initial population need to be generated. To this end, various initialization techniques are applicable, such as pseudo-random numbers, quasi-random sequences, and artificial immune systems [21].
- Subsequently, the fitness of each chromosome is evaluated to select the parent chromosomes to be paired. If the user requires to optimize multiple job scheduling performance measures, a multi-objective GA needs to be applied, and the crowding distance [22] of each chromosome should be measured.
- Two parental chromosomes are crossed over to generate offspring chromosomes that also undergo feasibility check. To this end, various crossover methods, including one-point crossover, two-point crossover, multiple-point crossover, uniform crossover, etc., are chosen from. Redundant or ineligible operations are randomly replaced by unassigned and eligible operations on randomly chosen machines [23]:

    (i) Redundant operations: The same operation has been performed on any machine.
    (ii) Ineligible operations: The previous operation has not been performed yet. Another ineligibility happens when the number of jobs to be supplied by a supplier exceeds the capacity of the supplier.

- Subsequently, a chromosome can be muted, which happens with a very low probability. This can be accomplished by changing the supplier of a job, changing the machine on which an operation is performed, etc. Chromosomes that have been mutated in specific ways may also need to be checked.
- The evolutional process will be terminated if [23]
  - (i) A predetermined number of generations or time has elapsed.
  - (ii) A satisfactory solution has been found.
  - (iii) No improvement in solution quality has taken place for a predetermined number of generations.

If the supplier and factory for each job have been determined, the problem in each factory becomes a flexible job shop scheduling problem.

2.5.3 XAI Applications for Visualizing Populations

Chromosome diagrams of different schedules also look similar. To address these issues, Wang and Chen [24] drew radar charts to represent chromosomes, thereby facilitating their discrimination, as shown in Fig. 2.10, which can be modified for a supply chain case by showing the supplier of each job, as shown in Fig. 2.11. In this figure, the raw materials for Job #1 are provided by Supplier #1, while the raw materials for the other jobs are provided by Supplier #2.

Another way to discriminate chromosomes is to use different colors for different jobs (see Fig. 2.12 a), resulting in color-encoded chromosomes [24]. Each color may appear many times for all operations of a job. In addition, to distinguish the operations assigned to different machines, a twin of the color-encoded chromosome can be drawn to display different machine numbers in gray with different shades (see Fig. 2.12 b).

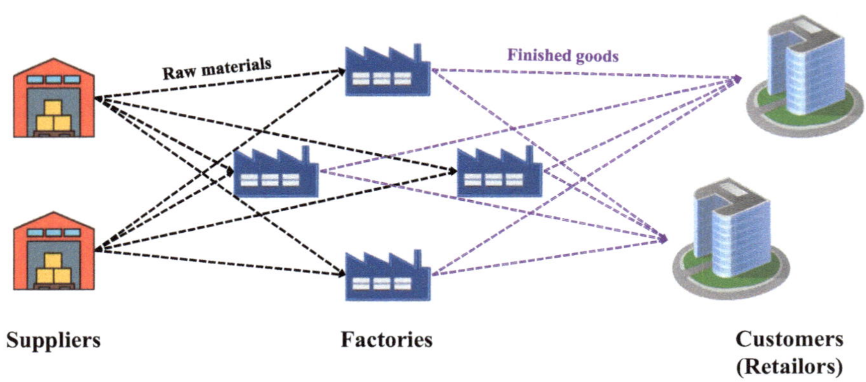

Fig. 2.7 Three-echelon supply chain

(1, 0, 1)	(2, 0, 2)	(3, 0, 1)

(1, 1, 3)	(1, 2, 1)	(2, 1, 3)	(2, 2, 4)	(3, 1, 3)	(1, 3, 2)	(3, 2, 1)	(2, 3, 4)

Fig. 2.8 Chromosome diagrams (or maps) for representing chromosome

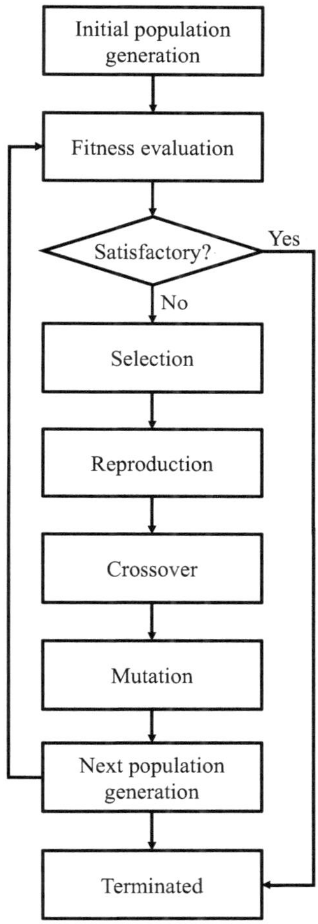

Fig. 2.9 Procedure for solving a scheduling problem using GA

If there are n jobs, n different colors should be chosen. These colors should be as different from each other as possible for users to distinguish them easily. To this end, a following mixed-integer nonlinear programming problem (MINP) can be solved [23].

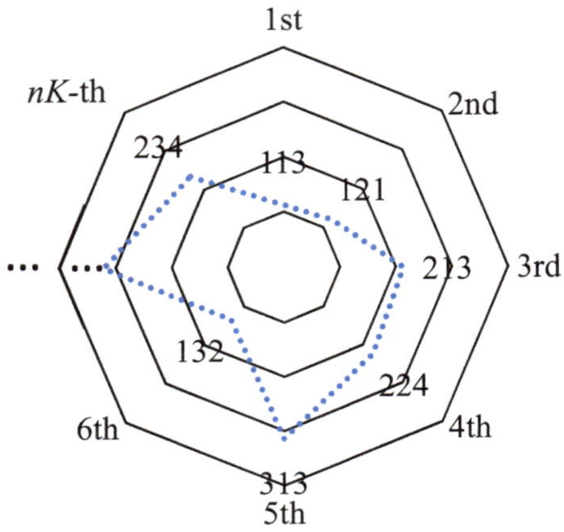

Fig. 2.10 Radar chart for visualizing chromosomes

However, it is difficult to overcome the scalability problem using chromosome diagrams. Therefore, a very wide chromosome map can be folded into a chromosome matrix (see Fig. 2.13) [23].

2.5.4 *XAI Applications for Visualizing the Evolution Process*

The crossover and mutation operations of GA can be visualized straightforward using color-encoded chromosome diagrams (see Fig. 2.14) [23].

The operations of GA can be visualized with decision trees (see Fig. 2.15) in which $\{\mathbf{x_s}\}$ is the set of chromosomes of the current population, $\mathbf{x_s} = \{x_{sl}\}$, and $x_{sl} = \{j_{sl}, o_{sl}, m_{sl}\}$. From $\{\mathbf{x_s}\}$, paired chromosomes $\{\mathbf{x_p}\}$ are selected.

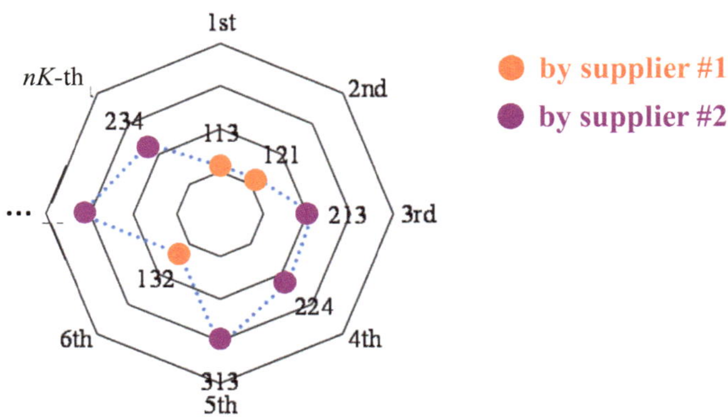

Fig. 2.11 Radar chart for visualizing chromosomes in supply chain cases

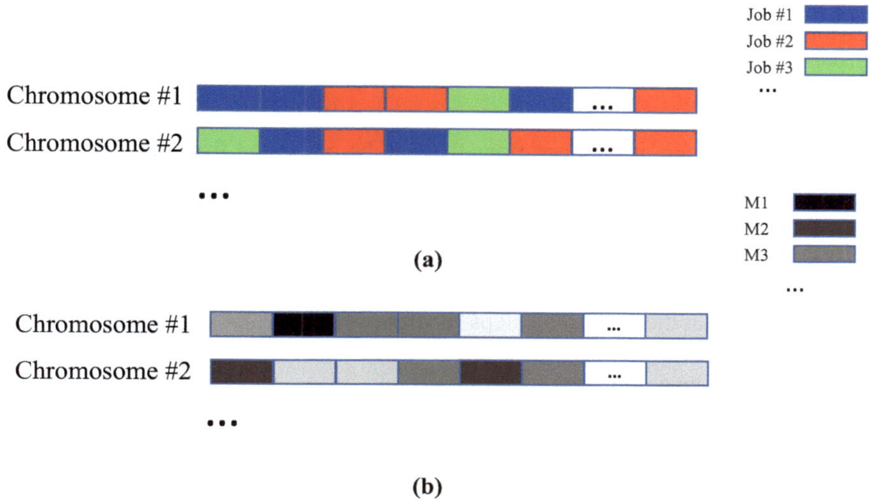

Fig. 2.12 Color coding of chromosomes

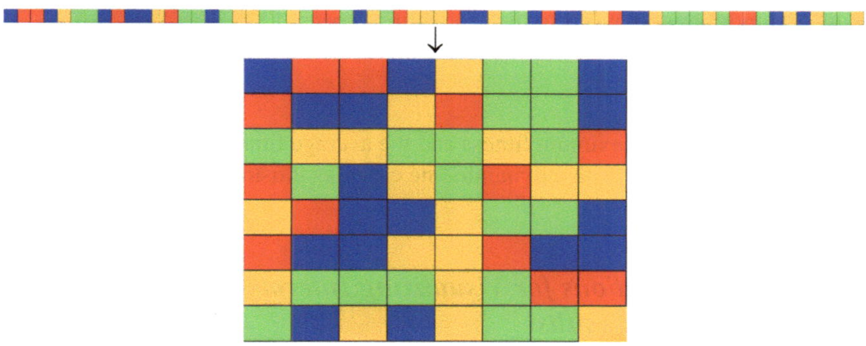

Fig. 2.13 Folding a chromosome map into a chromosomal matrix

A B&B algorithm is traditionally applied to find the optimal solution to a job scheduling problem [25], which forms a decision tree through branching and bounding to provide decision rules that can explain the scheduling mechanism (see Fig. 2.16) [26, 27]:

$$IF\ X_{113} = 1\ AND\ X_{121} = 1\ THEN\ Fitness \geq 13 \tag{2.52}$$

The B&B process is an iterative process, in which the lower bound on the fitness is estimated for each node, for which a lot of heuristics have been proposed [28, 29]. Then, the node with the highest lower bound is branched.

The B&B procedure is as follows:

Step 1. (Layer 0) Start from the root node *R*, which is usually the first operation of some job.

Step 2. $L = 1$.

Step 3. Select one of the operations that can be processed first; otherwise, go to Step 9.

Step 4. (Layer *L*) Process the operation on any available machine.

Step 5. (Layer *L*) Estimate the lower bound on the scheduling performance. For example,

(i). Assume that the remaining steps (operations) of all jobs can be processed on the fastest machine without delay.
(ii). Calculate the completion time of each job.
(iii). Find the maximum job completion time.
(iv). Calculate the completion time on each machine.
(v). Find the maximum machine completion time.
(vi). Evaluate the scheduling performance by the maximum of the results of (iii) and (v).

Step 6. (Layer *L*) Branch the node with the lowest lower bound, which is one of the most important operations to the scheduling process [30, 31].

Step 7. $L = L + 1$.

Step 8. Return to Step 3.

Step 9. Stop.

The changes in the maximum fitness and the average fitness during the evolution process are summarized with a dynamic line chart, as illustrated in Fig. 2.17.

2.5.5 XAI Applications for Visualizing Intermediate or Final Scheduling Results

The chromosomes of an intermediate population, including improved solutions, or the last population, including the best solutions, can be visualized using twin chromosome diagrams as well, as illustrated in Fig. 2.18, which reveals that it is better to process Jobs #1 or #2 earlier and Job #3 later. The distribution of machines is also visualized with the twin chromosome diagram, showing that machine M4 is more frequently in the intermediate population.

In addition, a scatter radar diagram can also be drawn to compare visually the chromosomes of an intermediate or the last population, as shown in Fig. 2.19. The evolution process of GA can be visualized by comparing the scatter radar diagrams of two consecutive populations, which is called a dynamic transition and contribution diagram.

Traditionally, final scheduling results can be represented by Gantt charts, disjunctive graphs, S-matrices (symbolic Gantt charts), Q-matrices, and others (see Fig. 2.20)

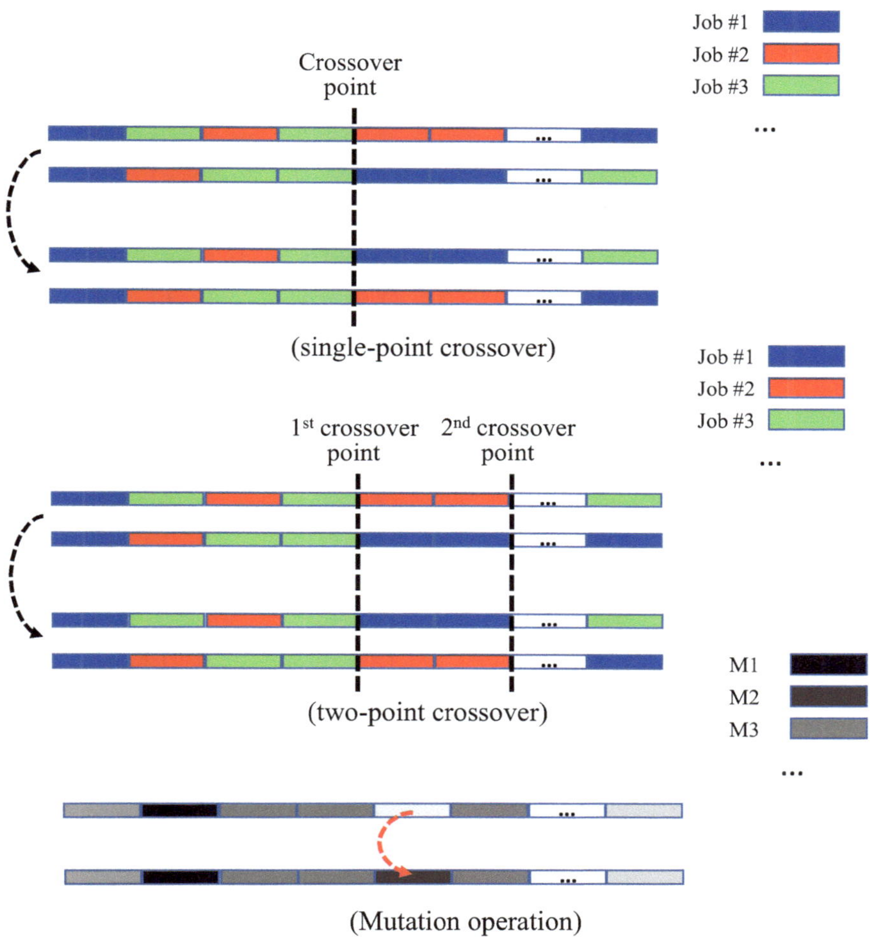

Fig. 2.14 Visualizing crossover and mutation operations of GA using color-encoded chromosome diagrams [23]

[32]. Among them, Q-matrices, showing the operations of each job, are less useful. Furthermore, the positions of nodes/operations in the disjunction graphs of two different chromosomes are the same, comparing their positions is not meaningful.

Color coding has been applied to discriminate between different jobs in Gantt charts [32], as shown in Fig. 2.21.

There is no need to design a twin color-encoded Gantt chart because the parts for different machines are already separated in the Gantt chart.

Chen [23] applied the concept of contrastive gradient-based saliency maps to the dynamic line chart to pinpoint the convergence point as soon as possible by associating points in the dynamic line chart with different colors to distinguish their

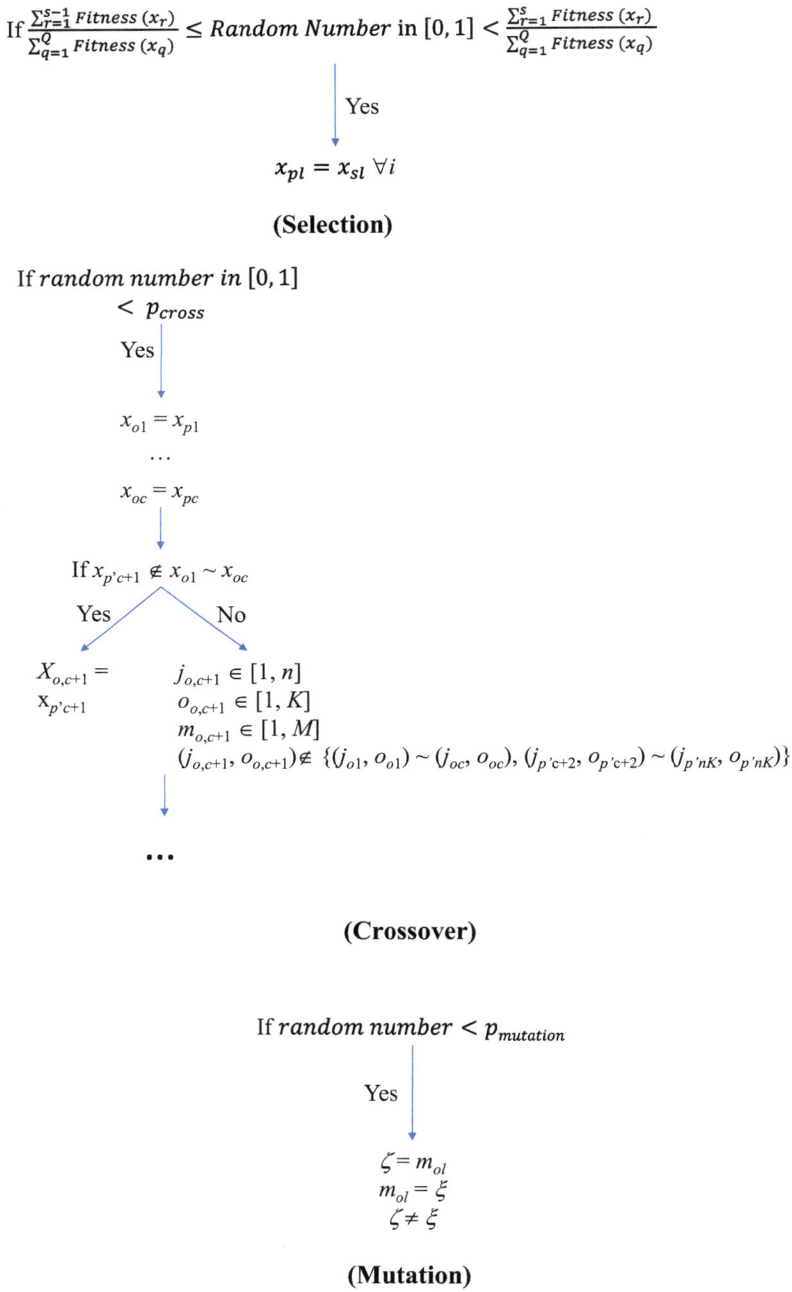

Fig. 2.15 Decision trees for explaining the operations of GA

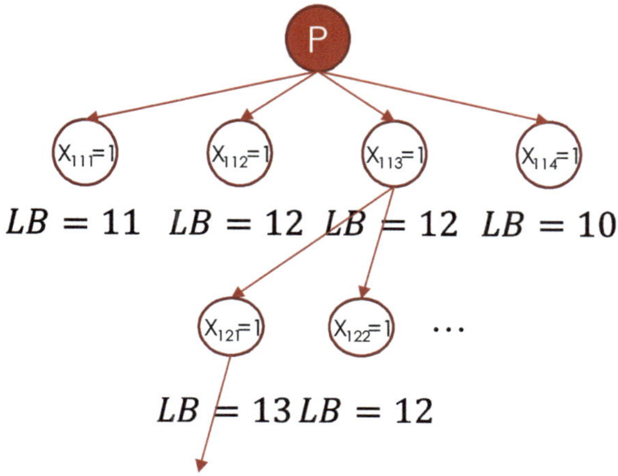

Fig. 2.16 Branch-and-bound-based decision tree

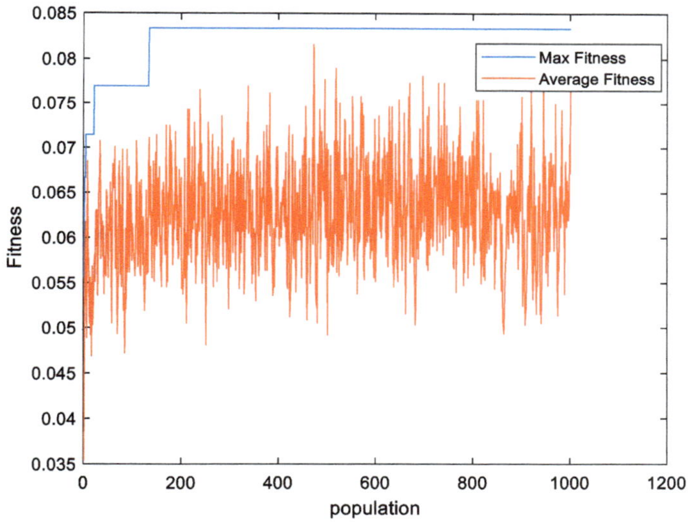

Fig. 2.17 Changes in the maximum fitness and the average fitness during the GA evolution process

performances:

$$R_{p_t} = 255 \cdot \left(\frac{\frac{p_{t+\Delta t} - p_t}{\Delta t} - \min\limits_{s}\left(\frac{p_{s+\Delta t} - p_s}{\Delta t}\right)}{\max\limits_{s}\left(\frac{p_{s+\Delta t} - p_s}{\Delta t}\right) - \min\limits_{s}\left(\frac{p_{s+\Delta t} - p_s}{\Delta t}\right)} \right)^{\varphi} \tag{2.53}$$

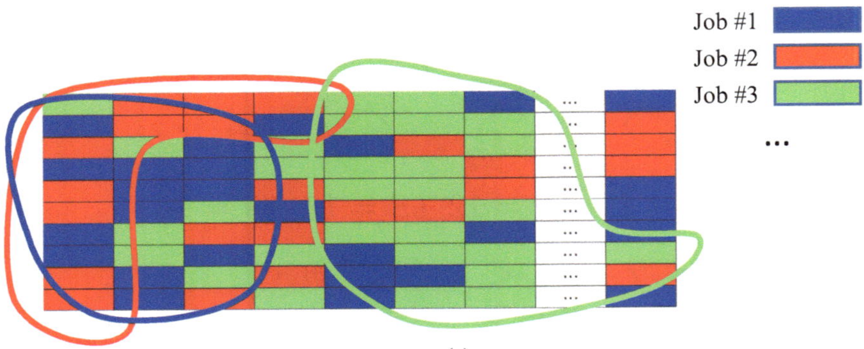

(a)

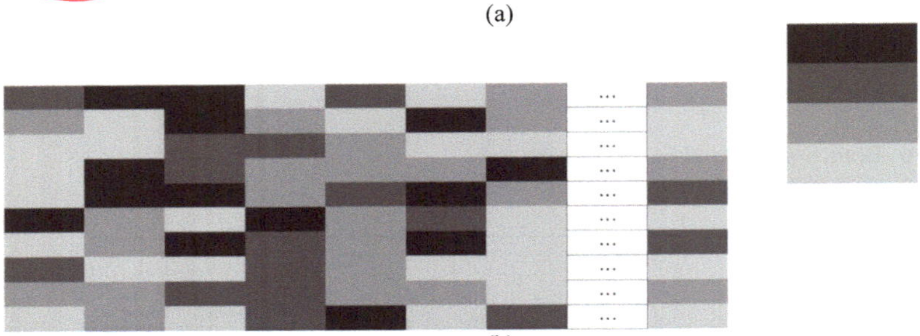

(b)

Fig. 2.18 Twin chromosome diagram used for illustrating an intermediate or the last population

$$G_{p_t} = 255 \cdot (1 - \left(\frac{\frac{p_{t+\Delta t} - p_t}{\Delta t} - \min_s\left(\frac{p_{s+\Delta t} - p_s}{\Delta t}\right)}{\max_s\left(\frac{p_{s+\Delta t} - p_s}{\Delta t}\right) - \min_s\left(\frac{p_{s+\Delta t} - p_s}{\Delta t}\right)} \right)^{\varphi}) \tag{2.54}$$

$$B_{p_t} = 0 \tag{2.55}$$

where p_t is the performance/fitness of population t and R_{p_t}, G_{p_t}, and B_{p_t} are the red, green, and blue color densities of (t, p_t). An example is given in Fig. 2.22, in which green points are associated with better scheduling performances than red points. Thus, when a data point turns green, it indicates that the performance is above the average. Whether the evolution process needs to be terminated depends on how green the current data point is for the scheduler.

Definition 2.1 The convergence point (p_{t_c}) is where all subsequent points are green, while all the points before that instant are not exactly green.

The goal is to detect the convergence point unambiguously and as fast as possible. Specifically,

$$\text{Min}\, p_{t_c} \tag{2.56}$$

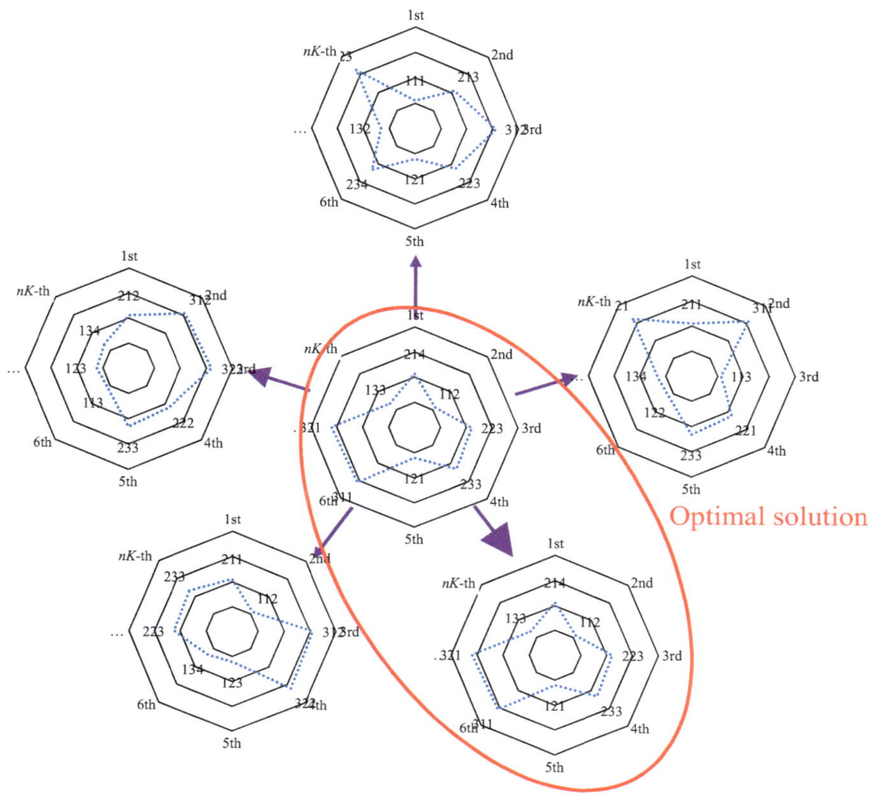

Fig. 2.19 Scatter radar diagram for visually comparing the chromosomes of the last population

s.t.

$$R_{p_t} \leq 0^+ \left(\text{or } G_{p_t} \geq 1^-\right) \forall t \geq t_c \tag{2.57}$$

$$R_{p_t}, G_{p_t} \in [0, 255] \forall p_t \tag{2.58}$$

$$p_t \in Z^+ \forall t \tag{2.59}$$

In addition, the applications of two popular XAI techniques, Shapley value (SHAP) analysis and local interpretable model-agnostic explanations (LIME) [33], in job scheduling for supply chains, are introduced in the next section.

2.5.6 SHAP and LIME Applications

In the case of artificial neural networks (ANNs) or deep neural networks (DNNs), the training results are the optimal settings of network parameters. However, in the case of GA, the training results are the modified inputs, which is more similar to the case of autoencoders. For this reason, SHAP and LIME may not be directly applicable to explain GA applications. Instead, a surrogate model, such as an ANN, classification and regression tree, random forest (RF), or extreme gradient boosting (XGBoost) [23, 33], is used to approximate the scheduling process and results of GA. Then SHAP and LIME are applied to the surrogate model.

A problem with this treatment is the low approximation accuracy using the surrogate model. Take the scheduling problem of a flexible job shop with three jobs, four machines, and eight operations in total as an example. The number of possible feasible solutions is up to $4^8 \cdot 8! = 2.64E+09$. The approximation accuracy using a DNN for GA with various numbers of examples is shown in Table 2.1:

$$MAE = \frac{\sum_{i=1}^{n} |y_i - \hat{y}_i|}{n} \tag{2.60}$$

$$RMSE = \sqrt{\frac{\sum_{i=1}^{n} (y_i - \hat{y}_i)^2}{n}} \tag{2.61}$$

$$R^2 = 1 - \frac{\sum_{i=1}^{n} (y_i - \hat{y}_i)^2}{\sum_{i=1}^{n} (y_i - \overline{y})^2} \tag{2.62}$$

which are summarized in Fig. 2.23. Obviously, the approximation accuracy improves rapidly with more trained examples.

To perform contribution analysis, the feasible solutions from all populations (see Fig. 2.24) can be input into XGBoost to predict their fitness values, and then SHAP

Table 2.1 Approximation accuracy of a DNN for the GA with various numbers of examples

	RMSE (Training)	MAE (Training)	R^2(Training)	RMSE (Training)	MAE (Training)	R^2(Training)
100	0.001	0.001	1.000	11.629	9.598	–17.865
1000	0.082	0.022	0.999	14.334	11.141	–31.754
2000	0.099	0.017	0.999	11.238	8.734	–19.089
3000	0.064	0.020	0.999	9.819	6.838	–14.234
4000	0.097	0.020	0.999	9.740	8.191	–13.555
5000	0.075	0.038	0.999	5.294	3.466	–3.015
6000	0.096	0.026	0.998	3.352	0.678	–0.588
7000	0.084	0.036	0.998	3.437	0.446	–0.539
8000	0.093	0.032	0.995	2.844	0.459	–0.153

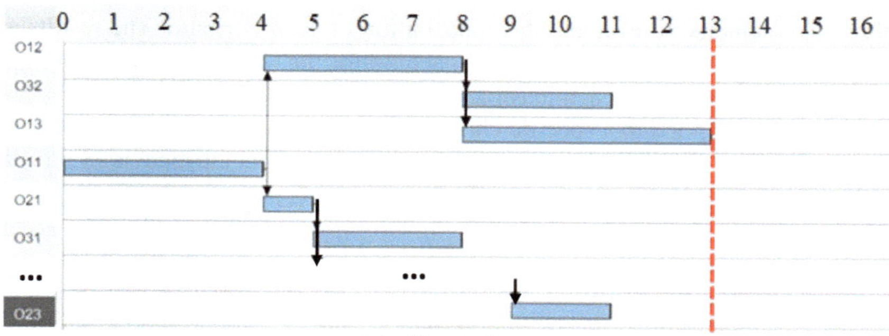

(a) Gantt chart

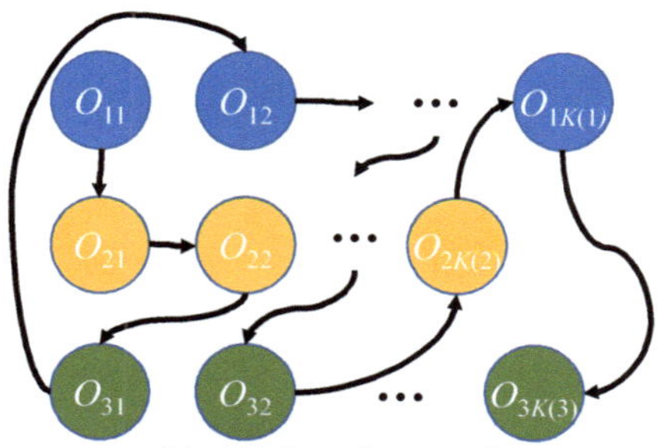

(b) Disjunctive graph

$$S = \begin{bmatrix} O_{12} & O_{32} & 0 & \dots & \\ O_{13} & 0 & \dots & & \\ O_{11} & O_{21} & O_{31} & 0 & \dots \\ \dots & & & & \\ O_{23} & 0 & \dots & & \end{bmatrix}$$

(c) S-matrix

$$Q = \begin{bmatrix} O_{11} & O_{12} & O_{13} & \dots & 0 \\ O_{21} & O_{22} & O_{23} & \dots & 0 \\ \dots & & & & \\ O_{31} & O_{32} & 0 & \dots & 0 \end{bmatrix}$$

(d) Q-matrix

Fig. 2.20 Traditional tools for illustrating the scheduling results

analysis is conducted to evaluate the contribution of each attribute. The results are shown in Fig. 2.25. Obviously, M7, the machine for the seventh operation was the most important attribute in each feasible solution.

LIME is also applied to a specific chromosome in the same case. The results are shown in Fig. 2.26. For this chromosome, the most important attribute is J8, i.e., the job number of the eighth operation in the chromosome.

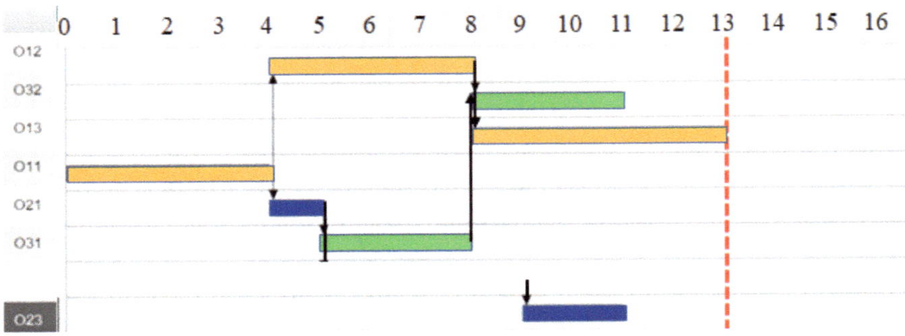

Fig. 2.21 Color coding used to discriminate between different jobs in a Gantt chart

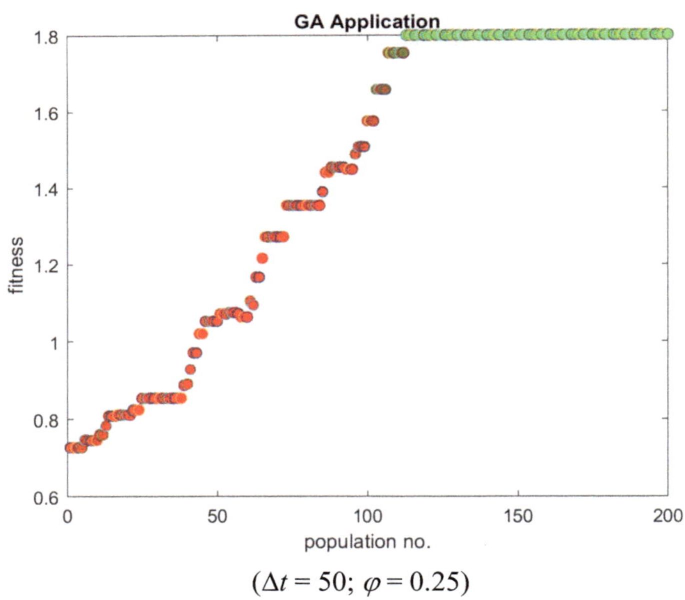

Fig. 2.22 Contrastive gradient-based saliency map application

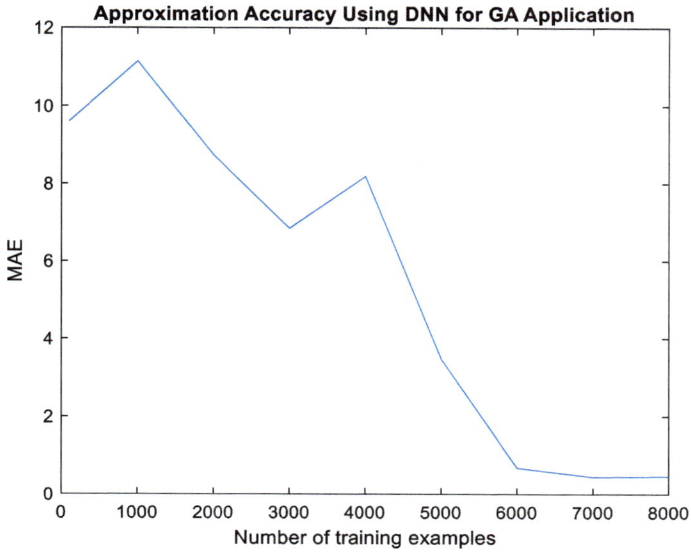

Fig. 2.23 Approximation accuracy using DNN for GA application

2.5.7 Color-Coded Pheromone Distribution Map

Chen et al. [27] designed a color-coded pheromone distribution map to show the scheduling process and results using ACO. In the map, ants' paths are shown within a rectangular area (see Fig. 2.27), and each row of the rectangle contains all possible operations. Each ant starts from the bottom row, selects one cell (operation) in each row, and continues until it reaches the top row to complete the entire path. Ants' accesses to different machines are represented by different colors.

During the scheduling process, the distribution of ants' visits on a color-coded pheromone distribution map changes continuously. For example, the color-coded pheromone distribution maps of the initial and final iterations are compared in Fig. 2.28.

Color-coded pheromone distribution maps can also be used to explain the scheduling process and results using another bio-inspired algorithm, such as GA, ABC, and PSO, for flexible job shops, supply chains, and others, as illustrated in Fig. 2.29.

	A	B	C	D	E	F	G	H	I	J	K	L	M	N	O	P	Q	R	S	T	U	V	W	X	Y	Z	AA	AB	AD
1	#	J1	O1	M1	J2	O2	M2	J3	O3	M3	J4	O4	M4	J5	O5	M5	J6	O6	M6	J7	O7	M7	J8	O8	M8	J9	O9	M9	Fitness
2	1	3	1	4	3	2	4	3	3	3	1	1	4	1	2	2	2	1	1	1	3	2	2	2	2	2	3	2	0.0256
3	2	2	1	1	1	1	3	2	2	4	1	2	3	2	3	3	1	3	1	3	1	3	3	2	3	3	3	3	0.0455
4	3	1	1	1	1	2	3	3	1	1	2	1	2	1	3	2	2	2	4	2	3	2	3	2	3	3	3	3	0.0417
5	4	2	1	1	2	2	3	1	1	3	2	3	1	3	1	3	1	2	2	1	3	2	3	2	3	3	3	4	0.0370
6	5	1	1	3	1	2	3	2	1	3	3	1	4	3	2	4	3	3	1	2	2	4	2	3	3	1	3	1	0.0556
7	6	1	1	1	2	1	2	2	2	1	1	2	2	3	1	4	2	3	4	1	3	3	3	2	1	3	3	2	0.0526
8	7	1	1	1	1	2	2	3	1	3	1	3	1	2	1	4	2	2	3	2	3	3	3	2	2	3	3	3	0.0417
9	8	3	1	3	2	1	1	2	2	1	1	1	4	2	3	1	3	2	3	1	2	1	3	3	3	1	3	4	0.0385
10	9	2	1	1	1	1	1	2	2	1	2	3	2	3	1	4	1	2	1	3	2	4	3	3	3	1	3	3	0.0476
11	10	3	1	2	2	1	1	3	2	3	2	2	4	1	1	4	2	3	4	3	3	2	1	2	2	1	3	4	0.0345
12	11	3	1	3	3	2	4	3	3	2	2	1	4	2	2	3	1	1	2	1	2	2	1	3	1	2	3	2	0.0345
13	12	3	1	2	3	2	3	2	1	3	3	3	1	2	2	2	1	1	4	1	2	1	1	3	3	2	3	3	0.0417
14	13	2	1	2	2	2	2	1	1	1	3	1	4	1	2	1	1	3	2	3	2	2	2	3	4	3	3	3	0.0476
15	14	1	1	3	1	2	4	2	1	3	1	3	2	3	1	1	3	2	1	3	3	4	2	2	4	2	3	2	0.0455
16	15	2	1	2	1	1	3	2	2	3	1	2	1	2	3	2	1	3	3	3	1	1	3	2	3	3	3	4	0.0400
17	16	3	1	2	2	1	4	3	2	2	2	2	4	3	3	1	2	3	4	1	1	3	1	2	3	1	3	1	0.0500
18	17	1	1	2	2	1	4	3	1	3	1	2	4	2	2	3	3	2	2	2	3	3	3	3	1	1	3	2	0.0588
19	18	3	1	2	3	2	3	3	3	3	1	1	3	2	1	2	1	2	2	2	2	2	2	3	3	1	3	1	0.0286

GAsummary +

Fig. 2.24 Feasible solutions from all iterations of the GA application

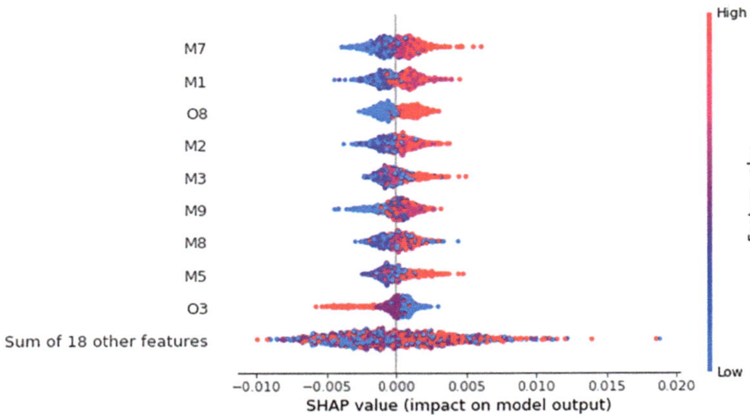

Fig. 2.25 SHAP analysis results (scatter plot)

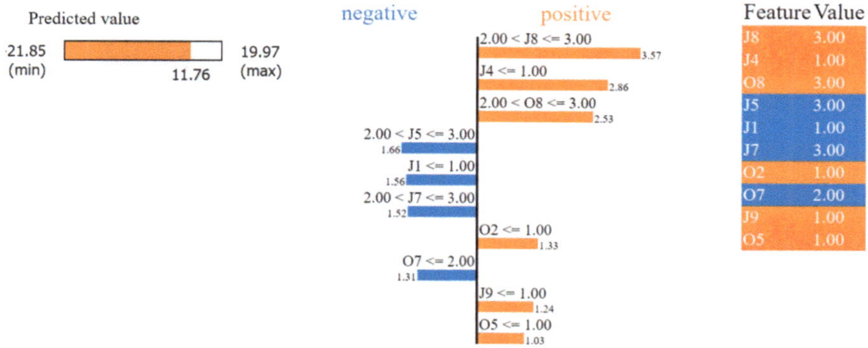

Fig. 2.26 LIME summary plot

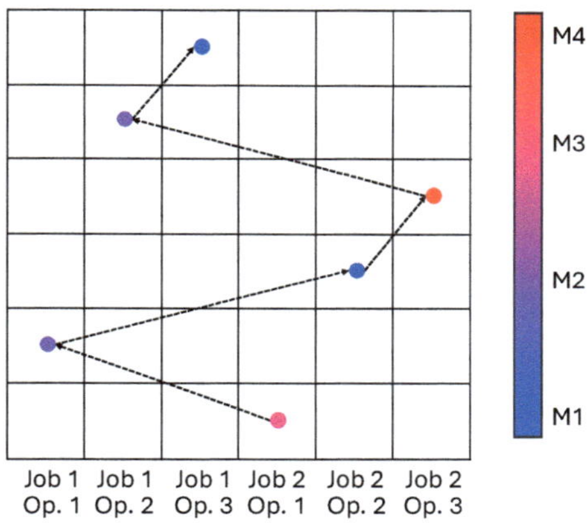

Fig. 2.27 Color-coded pheromone distribution map

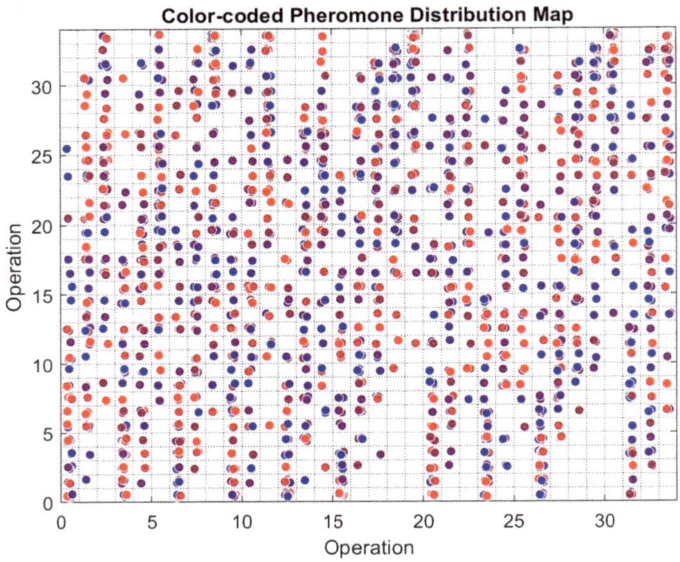

(a) Initial iteration

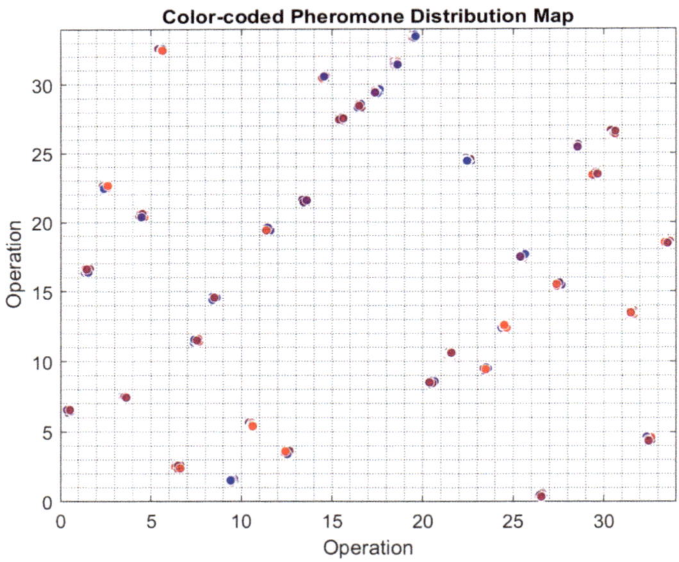

(b) Final iteration

Fig. 2.28 Comparison of the color-coded pheromone distribution maps of the initial and final iterations

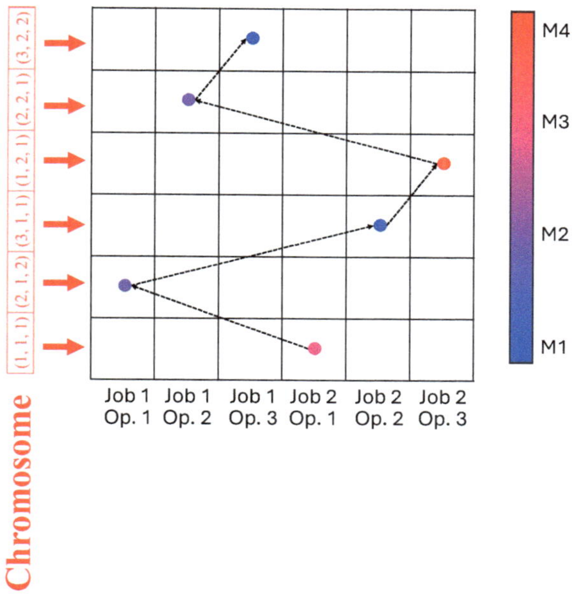

Fig. 2.29 Color-coded pheromone distribution map application to GA

References

1. K.Y. Liu, *Supply Chain Analytics: Concepts, Techniques and Applications* (Palgrave Macmillan, 2022)
2. S. Voß, D.L. Woodruff, *Introduction to Computational Optimization Models for Production Planning in a Supply Chain* (Springer, Berlin, 2006)
3. D. Ivanov, B. Sokolov, Dynamic supply chain scheduling. J. Sched. **15**(2), 201–216 (2012)
4. T.C.T. Chen, Cycle time prediction, in *Advances in Cycle Time Management* (2026), pp. 21–60
5. S.H. Zegordi, I.K. Abadi, M.B. Nia, A novel genetic algorithm for solving production and transportation scheduling in a two-stage supply chain. Comput. Ind. Eng. **58**(3), 373–381 (2010)
6. S.K. Jauhar, M. Pant, Genetic algorithms in supply chain management: a critical analysis of the literature. Sādhanā **41**(9), 993–1017 (2016)
7. M. Musavi, A. Bozorgi-Amiri, A multi-objective sustainable hub location-scheduling problem for perishable food supply chain. Comput. Ind. Eng. **113**, 766–778 (2017)
8. G. Wang, Integrated supply chain scheduling of procurement, production, and distribution under spillover effects. Comput. Oper. Res. **126**, 105105 (2021)
9. C. Pérez, L. Climent, G. Nicoló, A. Arbelaez, M.A. Salido, A hybrid metaheuristic with learning for a real supply chain scheduling problem. Eng. Appl. Artif. Intell. **126**, 107188 (2023)
10. A. Borumand, M.A. Beheshtinia, A developed genetic algorithm for solving the multi-objective supply chain scheduling problem. Kybernetes **47**(7), 1401–1419 (2018)
11. F. Xiong, P. Gong, P. Jin, J.F. Fan, Supply chain scheduling optimization based on genetic particle swarm optimization algorithm. Clust. Comput. **22**(Suppl 6), 14767–14775 (2019)
12. I. Dzalbs, T. Kalganova, Accelerating supply chains with ant colony optimization across a range of hardware solutions. Comput. Ind. Eng. **147**, 106610 (2020)
13. Y.C. Wang, T.C.T. Chen, C.W. Lin, Fuzzy reinforcement learning approach for supporting ubiquitous manufacturing facility selection in drone manufacturing. Int. J. Adv. Manuf. Technol. **141**(3), 1893–1903 (2025)

14. T.C.T. Chen, Fuzzy approach for production planning by using a three-dimensional printing-based ubiquitous manufacturing system. AI EDAM **33**(4), 458–468 (2019)
15. M. Kim, J. Park, Learning collaborative policies to solve np-hard routing problems. Adv. Neural. Inf. Process. Syst. **34**, 10418–10430 (2021)
16. A.D. Pia, S.S. Dey, M. Molinaro, Mixed-integer quadratic programming is in NP. Math. Program. **162**(1), 225–240 (2017)
17. Z.L. Chen, Integrated production and outbound distribution scheduling: review and extensions. Oper. Res. **58**(1), 130–148 (2010)
18. D. Oertel, R. Ravi, Complexity of transmission network expansion planning: NP-hardness of connected networks and MINLP evaluation. Energy Syst. **5**(1), 179–207 (2014)
19. T. Chen, M.C. Chiu, H.C. Wu, GenAI job scheduling system for solving a flexible job shop scheduling problem. AI EDAM **39**, e27 (2025)
20. T.C.T. Chen, Cycle time management, in *Advances in Cycle Time Management* (2026), pp. 1–20
21. V. Roshanaei, A. Azab, H. ElMaraghy, Mathematical modelling and a meta-heuristic for flexible job shop scheduling. Int. J. Prod. Res. **51**(20), 6247–6274 (2013)
22. S.B. Selçuklu, Multi-objective genetic algorithms, in *Handbook of Formal Optimization* (2024), pp. 1007–1044
23. T.C.T. Chen, Explaining genetic algorithm applications in job sequencing and scheduling, in *Explainable and Customizable Job Sequencing and Scheduling: Advancing Production Control and Management with XAI* (2025), pp. 73–102
24. Y.C. Wang, T. Chen, Adapted techniques of explainable artificial intelligence for explaining genetic algorithms on the example of job scheduling. Expert Syst. Appl. **237**(A), 121369 (2024)
25. C.P. Tomazella, M.S. Nagano, A comprehensive review of Branch-and-Bound algorithms: guidelines and directions for further research on the flowshop scheduling problem. Expert Syst. Appl. **158**, 113556 (2020)
26. M.C. Chiu, T.C.T. Chen, Y.C. Wang, Novel visualization XAI approach for assisting schedulers' comprehension and interpretation of ACO applications in job scheduling, in *2025 IEEE 15th Symposium on Computer Applications and Industrial Electronics* (2025), pp. 144–149
27. T.C.T. Chen, M.C. Chiu, Y.C. Lin, Novel visualization XAI techniques for assisting schedulers' comprehension and interpretation of ACO applications in job scheduling. Int. J. Hum.–Comput. Interact. 1–34 (2026)
28. D.R. Morrison, S.H. Jacobson, J.J. Sauppe, E.C. Sewell, Branch-and-bound algorithms: a survey of recent advances in searching, branching, and pruning. Discret. Optim. **19**, 79–102 (2016)
29. S. Du, J. Tong, W. Fan, Learning efficient branch-and-bound for solving mixed integer linear programs. Appl. Soft Comput. **172**, 112863 (2025)
30. M.C. Chiu, T.C.T. Chen, H.C. Wu, Enhancing the understandability and interpretability of type-II diabetes diagnosis using a deep fuzzy neural and XAI approach, in *2025 IEEE 15th Symposium on Computer Applications and Industrial Electronics* (2025), pp. 12–17
31. T.C.T. Chen, Management support, in *Advances in Cycle Time Management* (2026), pp. 97–135
32. N. Liu, Y.F. Zhang, W.F. Lu, Improving energy efficiency in discrete parts manufacturing system using an ultra-flexible job shop scheduling algorithm. Int. J. Precis. Eng. Manuf.-Green Technol. **6**, 349–365 (2019)
33. T.C.T. Chen, Explaining other bio-inspired algorithm applications in job sequencing and scheduling, in *Explainable and Customizable Job Sequencing and Scheduling: Advancing Production Control and Management with XAI* (2025), pp. 103–126

Chapter 3
XAI Applications in Facility Location Planning and Supply Network Optimization

3.1 Facility Location Planning and Supply Chain Management

Supply chain management (SCM) involves making many choices, including selecting upstream and downstream factories and suppliers, allocating orders among manufacturers, arranging logistics to minimize delays while maintaining minimal inventory, and so on. Among them, supplier selection is a critical task to the performance of supply chains. Many factors need to be considered when selecting suppliers, and these factors may not be in a single category. For example, Kilincci et al. [1] proposed 14 criteria for evaluating supplier suitability, which were divided into three categories: supplier performance, product performance, and service performance. When selecting suitable suppliers, these criteria can be considered in advance or dynamically as needed.

In addition to supplier selection, factory location (or facility sitting) is also crucial to enterprise development. Kahraman et al. [2] studied how automotive companies chose optimal factory locations. The criteria they considered included environmental regulations, community environment, competitive advantages, and political risks. Many studies have considered multi-level criteria; each level contains several sub-criteria, representing specific details of factory location. Appropriate site selection helps stabilize the supply chain and has a long-term impact on the development of related enterprises and the efficiency of the entire supply chain.

Previous literature has provided ample evidence that factory location is crucial for the sustainable development of a company. The principles of sustainable business operations must consider social, environmental, and economic dimensions. Therefore, multiple factors must be taken into account when selecting factory locations. When relocating a factory, planners not only focus on economic benefits but also consider social and environmental responsibilities to maximize overall well-being. Consequently, in SCM, social, environmental, and economic factors are considered simultaneously when developing the criteria for factory location evaluation [3].

T.-C. T. Chen, *Explainable Artificial Intelligence in Supply Chain Management*,
SpringerBriefs in Applied Sciences and Technology,
https://doi.org/10.1007/978-3-032-22564-1_3

Supplier selection and factory location selection are multi-criteria decision-making (MCDM) problems. Making a compromise among multiple criteria is a challenging task, for which a lot of artificial intelligence (AI) technologies have been applied.

3.2 Facility Location Planning in Semiconductor Supply Chain Localization

A special topic in facility location planning for supply chains is semiconductor supply chain localization.

In recent years, a trend in semiconductor supply chains has been to transfer semiconductor manufacturing capacity originally owned by wafer foundries back to the home countries of chip makers, the so-called localization [4]. The main driving force behind this phenomenon comes from the US–China trade war and regional political pressures that force chip designers to pursue wafer foundry capacity they can control [5]. In this way, the semiconductor supply chain can be more resilient, avoid war and political risks, and respond to rising environmental awareness [6, 7]. However, unlike the traditional facility location problem, selecting the suitable facility location for the localization required by customers is a novel issue that brings many challenges to wafer foundries [8]. More importantly, due to the sensitivity of this issue, it has not been fully addressed formally.

For instance, Wu and Jia [8], Reza-Gharehbagh et al. [9], and Chen [10] argued that semiconductor supply chain localization differs from traditional localization across various industries. For instance, semiconductor supply chain localization involves building new wafer fabs upstream, whereas traditional localization places new factories downstream. Moreover, the former is intended to mitigate geopolitical impacts [10, 11], while the latter aims to maximize profits [8, 9]. Consequently, existing supply chain localization methodologies and practices may lack relevance to the semiconductor context.

Some relevant literature is reviewed as follows. Belgacem and Murakami [12] argued that not only wafer fabrication plants (wafer fabs) but also the entire upstream and downstream semiconductor supply chain can be localized, making localized wafer fab location selection a more complex and integrated issue. So far, several backend packaging and testing companies have also decided to build new capacity near localized wafer fabs [13]. Additional impacts on the semiconductor supply chain need to be observed.

Merle [14] believed that bifurcation and localization are two recent trends in semiconductor supply chains. Bifurcation refers to the phenomenon that semiconductor supply chains can be roughly divided into supply chains controlled by the United States and China, respectively, while localization describes the current situation where most wafer fabs are concentrated in East Asia, Southeast Asia, and

Europe. However, the concept of bifurcation is closer to the localization happening recently.

In response to the growing trend of semiconductor supply chain localization, the European Union (EU) introduced the European Chip Act to foster a world-class chip ecosystem, enhance supply chain security, reduce reliance on external resources, and strengthen Europe's technological leadership [15]. However, directly competing with the US in developing new wafer fabrication facilities remains a major challenge. Thus, An and Yin [15] argued that the EU could reinforce its position in the upstream segments of the semiconductor supply chain and secure control over critical industry links by leveraging downstream demand and comparative advantages. This approach would bolster Europe's autonomy and resilience in the semiconductor sector.

3.3 AI Applications in Facility Location Planning for Supply Chains

Luthra et al. [16] emphasized the importance of greening and sustainability in establishing supply chains in supplier selection. Therefore, they proposed a systematic supplier selection evaluation system centered on sustainability from the perspective of organizational supply chains. Based on references and expert opinions, they determined 22 sustainable supplier selection criteria in three categories: economic, environmental, and social. They first used analytic hierarchy process (AHP) to determine the priorities of evaluation criteria, and then applied visekriterijumska optimizacija i kompromisno resenje (VIKOR) to screen suitable suppliers.

For supply chains in the food industry, Mohseni et al. [17] considered some criteria from three aspects (social, environmental, and economic) and various sub-criteria under these criteria to assist in plant site selection. They identified six drivers and seven barriers with the help of domain experts' opinions and applied several MCDM methods including AHP technique for order preference by similarity to ideal solution (TOPSIS), AHP, AHP-complex proportional assessment (COPRAS), Borda rule, and Copeland method to make a comprehensive decision.

Kheybari et al. [18] studied facility sitting for renewable energy production facilities. They considered the setting of a factory site from three aspects: social, environmental, and economic aspects, and established decision-making criteria for the three aspects. They derived the weight of criteria based on the data collected from an online survey using the best–worst method (BWM). The experimental results showed that energy saving, effect on resources, natural reserves, and wind flow were the most important factors for determining the locations of biogas, solar, and wind power plants.

Chen et al. [4] proposed the selectively calibrated derivation technique (SCDT)-generalized fuzzy technique for order preference by similarity to ideal solution (GFTOPSIS) method to assist wafer foundry companies in selecting suitable facility locations for customer-required capacity localization. Five criteria were formulated by them, including compliance with local needs, local government support and subsidies, total investment required, close relations with third parties, and low political and war risks. However, the SCDT-GFTOPSIS method may be difficult to understand and communicate, especially for users not familiar with fuzzy logic and TOPSIS.

Selecting an appropriate facility location based on the localization requirements of customers is clearly a fuzzy group decision-making (FGDM) process [19]. First, a group of decision-makers will be formed as this is critical to the sustainable development of a foundry. Additionally, there are many uncertainties due to the long planning horizon and the influence of external stakeholders such as local governments and customers. A number of artificial intelligence technologies have been applied to FGDM problems, including artificial neural networks (ANNs), genetic algorithms (GAs), support vector regression (SVR), ontology, etc. Decision-making is inherently an intuitive and straightforward process. However, after incorporating AI technologies, some FGDM processes become complicated and difficult to understand. Therefore, the need to apply such AI technologies is questionable [4].

Chen et al. [20] proposed an explainable fuzzy collaborative intelligence approach to prioritize candidate measures for addressing the challenges encountered by wafer foundries during semiconductor supply chain localization. In their framework, decision-makers applied various fuzzy MCDM methods based on their individual perspectives. The modified fuzzy weighted intersection was subsequently employed to aggregate their evaluations in a posteriori manner into a type-2 fuzzy approximation. To assess the importance of the criteria, fuzzy Shapley additive explanations analysis was performed. According to the experimental results, "early mass production" emerged as the best-performing measure, whereas "designing staggered shift systems" demonstrated the worst performance.

Chiu and Chen [21] introduced a neutrosophic collaborative intelligence approach to prioritize criteria for evaluating the effectiveness of measures aimed at mitigating the impacts of cultural barriers during the construction of localized wafer fabrication plants. In this approach, each expert applied the calibrated neutrosophic geometric mean to derive the neutrosophic weights of the criteria. The neutrosophic TOPSIS was subsequently applied to assess the suitability of each measure. Finally, the neutrosophic weighted intersection operator was developed to aggregate evaluation results from all experts. A systematic procedure was also established to guide measure prioritization. The indeterminacy of expert judgments was retained until the final stage, thereby enabling a flexible decision-making process.

Nafei et al. [22] applied neutrosophic AHP–TOPSIS to green supplier selection using single-valued neutrosophic sets (SVNSs). They reviewed six score functions to convert the neutrosophic judgment matrix into a crisp matrix. After applying all six score functions, the arithmetic mean and TOPSIS were used to rank multiple green suppliers. The ranks of each green supplier across these lists were aggregated using a sophisticated neural network.

Ubiquitous manufacturing systems is a special three-tier supply chain consisting of a design firm and multiple manufacturers, serving customers in virtually every part of the world. To plan the operations in the UM system, Wang et al. [23] proposed a fuzzy reinforcement learning approach. The fuzzy reinforcement learning approach employed the Q-learning algorithm to accelerate the prioritizing of criteria for 3D printing facility selection. Subsequent fuzzy technique for order preference by similarity to the ideal solution (FTOPSIS) application compared various 3D printing facilities to ensure that drone orders are correctly assigned to suitable UM partners.

3.4 XAI Applications in Facility Location Planning for Supply Chains

Wang et al. [23] proposed a fuzzy reinforcement learning approach to evaluate the priorities of criteria for choosing suitable 3D printing facilities. In the fuzzy reinforcement learning approach, first the supply chain planner compares the relative priority of criterion i over that of criterion j, denoted by $\tilde{a}_{ij}$ using linguistic terms such as "slightly more important than" and "extremely more important than"; $i, j \in [1, n]$. Such linguistic terms can be expressed in triangular fuzzy numbers (TFNs) from 1 to 9. $\tilde{\mathbf{A}} = [\tilde{a}_{ij}]$.

The fuzzy eigenvalue and eigenvector of $\tilde{\mathbf{A}}$, indicated with $\tilde{\lambda}$ and $\tilde{\mathbf{x}}$, respectively, can be derived by

$$\det\left(\tilde{\mathbf{A}}(-)\tilde{\lambda}\mathbf{I}\right) = 0 \tag{3.1}$$

and

$$\left(\tilde{\mathbf{A}}(-)\tilde{\lambda}\mathbf{I}\right)(\times)\tilde{\mathbf{x}} = 0 \tag{3.2}$$

in which (–) and (×) are fuzzy subtraction and multiplication operators, respectively. There are at most n solutions to the two equations. Among them, the maximum value of $\tilde{\lambda}$ is indicated with $\tilde{\lambda}_{\max}$, where the associated fuzzy eigenvector is $\tilde{\mathbf{x}}_{\max}$, based on which the consistency among judgments can be evaluated with fuzzy consistency ratio $\widetilde{CR}$ as

$$\widetilde{CR} = \frac{\frac{\tilde{\lambda}_{\max} - n}{n-1}}{RI} \tag{3.3}$$

where RI is the random consistency index. For these judgments to be consistent, $\widetilde{CR}$ has to be lower than 0.1. In addition, the fuzzy priorities of criteria are derived as

$$\tilde{\mathbf{w}} = [\tilde{w}_i] = \frac{\tilde{\mathbf{x}}_{\max}}{|\tilde{\mathbf{x}}_{\max}|} \tag{3.4}$$

However, solving Eqs. (3.1) and (3.2) is a challenging task due to the involvement of fuzzy multiplication. Most past studies applied approximation methods. However, we believe that only when the fuzzy priorities of criteria can be exactly derived can the final decision not be misled. Therefore, ACO is applied to derive the values of $\tilde{\lambda}$ and $\tilde{\mathbf{x}}$ exactly as follows.

First, fuzzy parameters and variables in Eqs. (3.1) and (3.2) are replaced with their α cuts:

$$\det\left(\tilde{\mathbf{A}}(\alpha)(-)\tilde{\lambda}(\alpha)\mathbf{I}\right) = 0 \tag{3.5}$$

$$\left(\tilde{\mathbf{A}}(\alpha)(-)\tilde{\lambda}(\alpha)\mathbf{I}\right)(\times)\tilde{\mathbf{x}}(\alpha) = 0 \tag{3.6}$$

If α takes 11 possible values (0, 0.1, …, 1), Eqs. (3.5) and (3.6) need to be solved $11{\cdot}2^{C_2^n}$ times to derive the α cuts of fuzzy maximal eigenvalue and fuzzy eigenvector as

$$\lambda^L(\alpha) = \min_{\det\left(\left[a_{ij}^*(\alpha)\right]-\lambda_t(\alpha)\mathbf{I}\right)=0} (\lambda_t(\alpha)) \tag{3.7}$$

$$\lambda^R(\alpha) = \max_{\det\left(\left[a_{ij}^*(\alpha)\right]-\lambda_t(\alpha)\mathbf{I}\right)=0} (\lambda_t(\alpha)) \tag{3.8}$$

$$\mathbf{x}^L(\alpha) = \min_{([a_{ij}^*(\alpha)]-\lambda_t(\alpha)\mathbf{I})\mathbf{x}_t(\alpha)=0} (\mathbf{x}_t(\alpha)) \tag{3.9}$$

$$\mathbf{x}^R(\alpha) = \max_{([a_{ij}^*(\alpha)]-\lambda_t(\alpha)\mathbf{I})\mathbf{x}_t(\alpha)=0} (\mathbf{x}_t(\alpha)) \tag{3.10}$$

where $* = L$ or R. $\lambda_t^L(\alpha)$, $\lambda_t^R(\alpha)$, $\mathbf{x}_t^L(\alpha)$, and $\mathbf{x}_t^R(\alpha)$ are the results derived from the tth combination; $t = 1 \sim 11{\cdot}2^{C_2^n}$. To enhance the efficiency of ACO, a Q-learning algorithm is designed as follows:

Step 1. Initialize the $Q(s, a)$ value functions, which can be done arbitrarily.

Step 2. Perceive the current state s_0.

Step 3. Following a predetermined learning policy, such as the ε (epsilon)–greedy policy, to select an appropriate action (a) for the given state (s_0):ε is usually set to a small value with a probability of $1 - \varepsilon$ to use argmax (i.e., choosing the most ideal behavior shown in the Q-table) and a probability of ε to use random behavior (i.e., completely ignoring the results of the Q-table).

Step 4. Execute the selected action (a), receive the immediate reward (r), and perceive the next state s_1.

Step 5. Update the value function as follows:

$$Q(s_0, a) = Q(s_0, a) + \alpha[r + \gamma \max_b Q(s_1, b) - Q(s_0, a)] \tag{3.11}$$

where α, the step size parameter, determines the learning rate; γ, the discount-rate parameter, $0 \leq \gamma \leq 1$, impacts the present value of future rewards. If the drone manufacturer does not prefer any action for a specific state, all the $Q(s, a)$ values in the policy table can be initialized with the same value.

Step 6. Let $s_0 = s_1$.

Step 7. Go to Step 3 until state s_0 represents a terminal state.

Step 8. Repeat Steps 2 to 7 if the maximum number of episodes has not been completed.

To support ACO, in the Q-learning algorithm, the objective is to maximize the ranges of α cuts. When a new α cut of some judgment is taken into account, an eigenanalysis is conducted to derive the α cuts of the fuzzy priorities of all criteria, i.e., $w_{i,t}^*(\alpha)$, which are compared to the minimum and maximum values, i.e., $\min_{u<t} w_{i,u}^L(\alpha)$ and $\max_{u<t} w_{i,u}^R(\alpha)$ so far. If $w_{i,t}^*(\alpha) < \min_{u<t} w_{i,u}^L(\alpha)$ or $w_{i,t}^*(\alpha) > \max_{u<t} w_{i,u}^R(\alpha)\ \forall i$, the learning machine agent receives an additional reward of $+1$. Otherwise, the learning agent receives an additional reward of -1, i.e., a penalty. It is worth noting that the learning machine agent can receive at most $+n$ additional rewards, or $-n$ additional penalties.

Chen et al. [4] proposed the SCDT-GFTOPSIS method for selecting suitable facility locations in customer-required capacity localization. In SCDT, each decision-maker constructs a fuzzy judgment matrix $\tilde{\mathbf{A}}(m)$. $\tilde{a}_{ij}(m)$ is the relative priority of criterion i over criterion j compared by decision-maker m; $i, j = [1, n]$; $m = 1 \sim M$. Then, fuzzy geometric mean (FGM) [24] and fuzzy inverse column sum method (FICSM) [25] are combined to derive the fuzzy priorities of criteria:

(Scdt)

Step 1. Derive the fuzzy priority of criterion i using FGM as [24] follows:

$$\tilde{w}_i(FGM) \cong \frac{\sqrt[n]{\prod_{j=1}^{n} \tilde{a}_{ij}}}{\sum_{k=1}^{n} \sqrt[n]{\prod_{j=1}^{n} \tilde{a}_{kj}}}; i = 1 \sim n \tag{3.12}$$

Namely,

$$w_{i1}(FGM) \cong \frac{1}{1 + \sum_{m \neq i} \frac{\sqrt[n]{\prod_{j=1}^{n} a_{mj3}}}{\sqrt[n]{\prod_{j=1}^{n} a_{ij1}}}}; i = 1 \sim n \tag{3.13}$$

$$w_{i2}(FGM) \cong \frac{1}{1 + \sum_{m \neq i} \frac{\sqrt[n]{\prod_{j=1}^{n} a_{mj2}}}{\sqrt[n]{\prod_{j=1}^{n} a_{ij2}}}}; i = 1 \sim n \tag{3.14}$$

$$w_{i3}(FGM) \cong \frac{1}{1 + \sum_{m \neq i} \frac{\sqrt[n]{\prod_{j=1}^{n} a_{mj1}}}{\sqrt[n]{\prod_{j=1}^{n} a_{ij3}}}}; i = 1 \sim n \tag{3.15}$$

Step 2. Derive the fuzzy priority of criterion i using FICSM as [25] follows:

$$\tilde{w}_i(FICSM) \cong \frac{\frac{1}{\sum_{m=1}^{n} \tilde{a}_{mi}}}{\sum_{j=1}^{n} \frac{1}{\sum_{m=1}^{n} \tilde{a}_{mj}}}; i = 1 \sim n \tag{3.16}$$

Namely,

$$w_{i1}(FICSM) \cong \frac{1}{1 + \sum_{j \neq i} \frac{\sum_{m=1}^{n} a_{mi3}}{\sum_{m=1}^{n} a_{mj1}}}; i = 1 \sim n \tag{3.17}$$

$$w_{i2}(FICSM) \cong \frac{1}{1 + \sum_{j \neq i} \frac{\sum_{m=1}^{n} a_{mi2}}{\sum_{m=1}^{n} a_{mj2}}}; i = 1 \sim n \tag{3.18}$$

$$w_{i3}(FICSM) \cong \frac{1}{1 + \sum_{j \neq i} \frac{\sum_{m=1}^{n} a_{mi1}}{\sum_{m=1}^{n} a_{mj3}}}; i = 1 \sim n \tag{3.19}$$

Step 3. Derive the crisp priority of criterion i as $w_i(CRISP)$.

Step 4. If $\sum_{i=1}^{n} |w_{i2}(FICSM) - w_i(CRISP)| \leq \sum_{i=1}^{n} |w_{i2}(FGM) - w_i(CRISP)|$, derive the fuzzy priority of criterion i using SCDT as

$$\tilde{w}_i(SCDT) \cong \tilde{w}_i(FICSM) \cdot \frac{w_i(CRISP)}{w_{i2}(FICSM)}; i = 1 \sim n \tag{3.20}$$

Namely,

$$w_{i1}(SCDT) \cong w_{i1}(FICSM) \cdot \frac{w_i(CRISP)}{w_{i2}(FICSM)}; i = 1 \sim n \tag{3.21}$$

$$w_{i2}(SCDT) \cong w_{i2}(FICSM) \cdot \frac{w_i(CRISP)}{w_{i2}(FICSM)}; i = 1 \sim n \tag{3.22}$$

$$w_{i3}(SCDT) \cong w_{i3}(FICSM) \cdot \frac{w_i(CRISP)}{w_{i2}(FICSM)}; i = 1 \sim n \tag{3.23}$$

Otherwise,

$$\tilde{w}_i(SCDT) \cong \tilde{w}_i(FGM) \cdot \frac{w_i(CRISP)}{w_{i2}(FGM)}; i = 1 \sim n \tag{3.24}$$

Namely,

$$w_{i1}(SCDT) \cong w_{i1}(FGM) \cdot \frac{w_i(CRISP)}{w_{i2}(FGM)}; i = 1 \sim n \tag{3.25}$$

$$w_{i2}(SCDT) \cong w_{i2}(FGM) \cdot \frac{w_i(CRISP)}{w_{i2}(FGM)}; i = 1 \sim n \tag{3.26}$$

$$w_{i3}(SCDT) \cong w_{i3}(FGM) \cdot \frac{w_i(CRISP)}{w_{i2}(FGM)}; i = 1 \sim n \tag{3.27}$$

SCDT is subject to low understandability and transparency because it derives the fuzzy priorities of criteria by solving a fuzzy eigenanalysis problem [26, 27], the reasoning mechanism of which is complex and difficult to reveal in details.

3.4.1 *Interpreting the Fuzzy Priority Derivation of Criteria*

The understandability of priority evaluation results can be improved by better visualization of pairwise comparison inputs. According to Lin and Chen [28], pairwise comparison results can be summarized by a gradient bar chart for the decision-maker to confirm, as shown in Fig. 3.1. According to this figure, $\tilde{a}_{21}(m) = \left(\frac{1}{3}, 1, 1\right)$, i.e., "as equal as", and $\tilde{a}_{31}(m) = (3, 5, 7)$.

The effectiveness of this treatment can be assessed by how accurately a viewer estimates the relative priorities from the bars:

$$\widetilde{SAD} = \sum_{i,j} \left| \hat{\tilde{a}}_{ij}(m)(-)\tilde{a}_{ij}(m) \right| \tag{3.28}$$

where $\widetilde{SAD}$ is the fuzzy sum of absolute deviations and $\hat{\tilde{a}}_{ij}(m)$ is the estimated value of $\tilde{a}_{ij}(m)$ after viewing Fig. 3.1. To facilitate the calculation, $\widetilde{SAD}$ is defuzzified as [29]

$$SAD = \sum_{i,j} \left(\left| \hat{a}_{ij1}(m) - a_{ij1}(m) \right| + 2\left| \hat{a}_{ij2}(m) - a_{ij2}(m) \right| + \left| \hat{a}_{ij3}(m) - a_{ij3}(m) \right| \right) \tag{3.29}$$

SAD is the smaller the better, i.e., the smaller the *SAD*, the higher the understandability of SCDT. For example, a viewer estimated the relative priorities to be (1, 1, 1), (0, 0.5, 1), and (0, 2, 7) after viewing Fig. 3.1. Then,

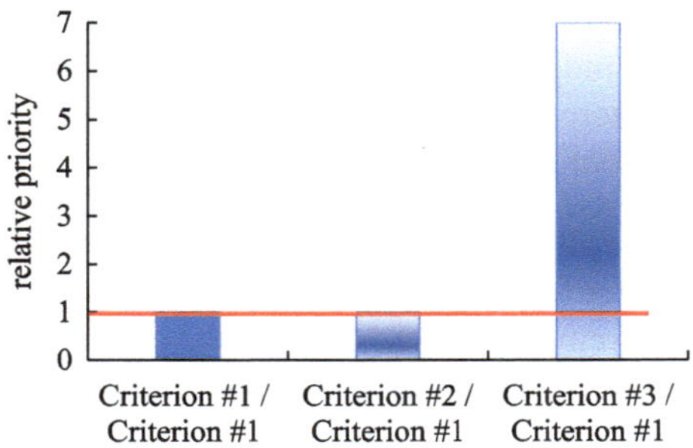

Fig. 3.1 Gradient bar chart for the relative priorities of Criterion #1 with respect to other criteria

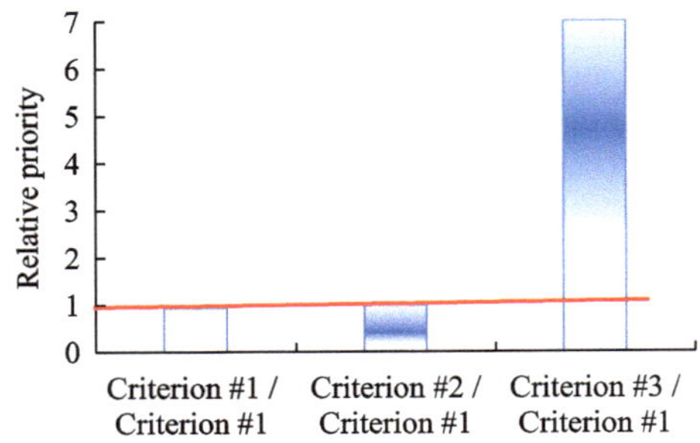

Fig. 3.2 Hanging gradient bar chart for the relative priorities of Criterion #1 with respect to other criteria

$$\widetilde{SAD} = |(1,1,1)(-)(1,1,1)|(+)\left|(0,0.5,1)(-)\left(\frac{1}{3},1,1\right)\right|(+)|(0,2,7)(-)(3,5,7)|$$

and

$$SAD = 0 + \frac{1}{3} + 3 + 2 \cdot (0 + 0.5 + 3) + 0 + 0 + 0 = 10.333$$

Figure 3.1 is not precise because each fuzzy priority started from 0. To accurately represent these fuzzy numbers with non-zero lower bounds, the bars may be left hanging (see Fig. 3.2). It is worth considering whether to change to a horizontal bar chart, such as a force plot for SHAP analysis.

After viewing Fig. 3.2, the viewer estimated the relative priorities to be 1, (0.25, 0.5, 1), and (3, 5, 7). Then,

$$\widetilde{SAD} = |(1,1,1)(-)(1,1,1)|(+)\left|(0.25,0.5,1)(-)\left(\frac{1}{3},1,1\right)\right|(+)|(3,5,7)(-)(3,5,7)|$$

and

$$SAD = 0 + 0.083 + 0 + 2 \cdot (0 + 0.5 + 0) + 0 + 0 + 0 = 1.083$$

which is much better than the result after viewing Fig. 3.1.

A judgment matrix can be illustrated using a heatmap, as illustrated in Fig. 3.3.

Therefore, theoretically, a fuzzy judgment matrix can be illustrated using a gradient heatmap, as illustrated in Fig. 3.4.

Example 3.1 The case discussed by Chen et al. [4] is used as an example. In this case, a wafer foundry selects a facility location to build a new wafer fab to meet customer requirements. Three domain experts form a team to compare several alternatives and recommend the most appropriate facility location to the company. Five criteria are formulated, including compliance with local needs, local government support and subsidies, total investment required, close relations with third parties, and low political and war risks. Table 3.1 describes the rationale for the selection and the validation methods of the five criteria.

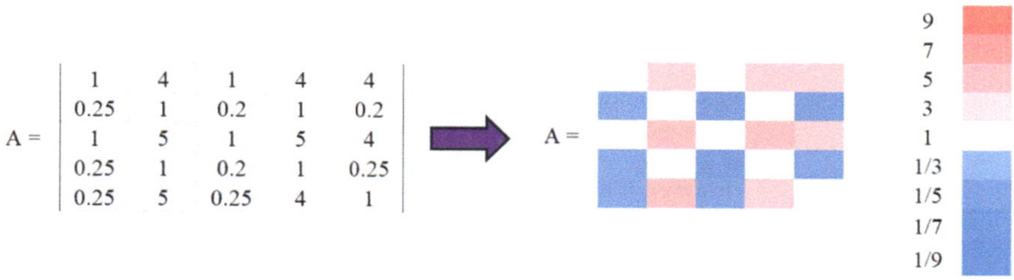

Fig. 3.3 Heatmap for illustrating a judgment matrix

Fig. 3.4 Gradient heatmap for illustrating a fuzzy judgment matrix

Take the first decision-maker as an example. The pairwise comparison results by the decision-maker are summarized by the following fuzzy judgment matrix:

$$\tilde{\mathbf{A}}(1) = \begin{bmatrix} 1 & (1,\ 3,\ 5) & 1/(1,\ 3,\ 5) & (3,\ 5,\ 7) & (3,\ 5,\ 7) \\ 1/(1,\ 3,\ 5) & 1 & 1/(1,\ 3,\ 5) & 1/(1,\ 3,\ 5) & (2,\ 4,\ 6) \\ (1,\ 3,\ 5) & (1,\ 3,\ 5) & 1 & (2,\ 4,\ 6) & (3,\ 5,\ 7) \\ 1/(3,\ 5,\ 7) & (1,\ 3,\ 5) & 1/(2,\ 4,\ 6) & 1 & (3,\ 5,\ 7) \\ 1/(3,\ 5,\ 7) & 1/(2,\ 4,\ 6) & 1/(3,\ 5,\ 7) & 1/(3,\ 5,\ 7) & 1 \end{bmatrix}$$

After applying the proposed methodology, a hanging gradient bar chart is drawn to illustrate the pairwise comparison results for the decision-maker to confirm (see Fig. 3.5). The effectiveness of the hanging gradient bar chart in improving the understandability of SCDT is assessing using $\widetilde{SAD}$. The results are summarized in Table 3.2.

Let $\tilde{w}_i(m)$ be the fuzzy priority of criterion i for decision-maker m. The value of $\tilde{w}_i(m)$ can be derived using a fuzzy eigenanalysis:

$$\det\left(\tilde{\mathbf{A}}(m)(-)\tilde{\lambda}(m)\mathbf{I}\right) = 0 \tag{3.30}$$

$$\left(\tilde{\mathbf{A}}(m)(-)\tilde{\lambda}(m)\right)(\times)\tilde{\mathbf{x}}(m) = 0 \tag{3.31}$$

Table 3.1 Selection and validation of the five criteria

Criterion	Selection rationale	Validation method
Compliance with local needs	• The technologies used in new fabs should be more advanced than those of existing competitors, such as 3–5 nm • However, if the product demand of mature technologies has not been fully met, it is also feasible to build new fabs with older technologies (such as 7–28 nm) • Customer pressure for localized wafer fabrication capacity is also one of these demands	Statistics on wafer fabrication technology requirements for local customers' products
Local government support and subsidies	• Support and subsidies from local governments, including tax exemptions, land acquisition assistance, direct subsidies, improvements to public facilities such as water, electricity, and transportation, and assistance in recruiting and training labor • The local governments that the company actually considered all provided financial support	Negotiation results of tax incentives and breaks with local authorities
Total investment required	• The total investment required was approximately US$10–50 billion	Learning from the experience of building the first new fab
Close relations with third parties	• Need to closely connect with third parties in the global semiconductor supply chain (such as raw material or equipment suppliers, downstream manufacturers, etc.)	The current and future number of third parties in the region
Low political and war risks	• Political, war, and other (such as financial and epidemic) risks need to be low	Political stability index[a]

[a]https://www.theglobaleconomy.com/rankings/wb_political_stability/

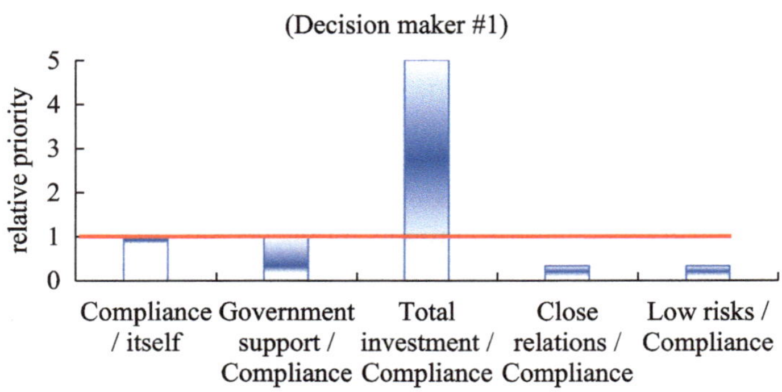

Fig. 3.5 Hanging gradient bar chart for the relative priorities of compliance with respect to other criteria

Table 3.2 Effectiveness of the hanging gradient bar chart in improving the understandability

Figure	Bar chart	Gradient bar chart	Hanging gradient bar chart
SAD	13.28	7.65	1.33
Improvement %	–	42%	90%

$$\tilde{w}_i(m) = \frac{\tilde{x}_i(m)}{\sum_{j=1}^{n} \tilde{w}_j(m)} \tag{3.32}$$

where $\tilde{\lambda}(m)$ and $\tilde{\mathbf{x}}(m)$ represent the fuzzy eigenvalue and eigenvector of $\tilde{\mathbf{A}}(m)$; (–) and ($\times$) denote fuzzy subtraction and multiplication, respectively. According to Lin and Chen [27], the results can also be summarized by a gradient bar chart, as shown in Fig. 3.6, which precision is improved by devising a hanging gradient bar chart instead (see Fig. 3.7).

In Fig. 3.6, it is noteworthy that the lower bound and core of $\tilde{w}_1(m)$ are the same. Furthermore, the cores, not the upper bounds, of all fuzzy priorities add up to 1. The effectiveness of Figs. 3.5 and 3.6 can also be evaluated and compared using Eq. (3.28).

Hanging gradient bar charts help to check the correctness of the (relative) priorities of criteria and allow comparisons between various derivation methods. If SCDT outperforms other derivation methods, its hanging gradient bar charts will be most

Fig. 3.6 Gradient bar chart for the fuzzy priorities of criteria

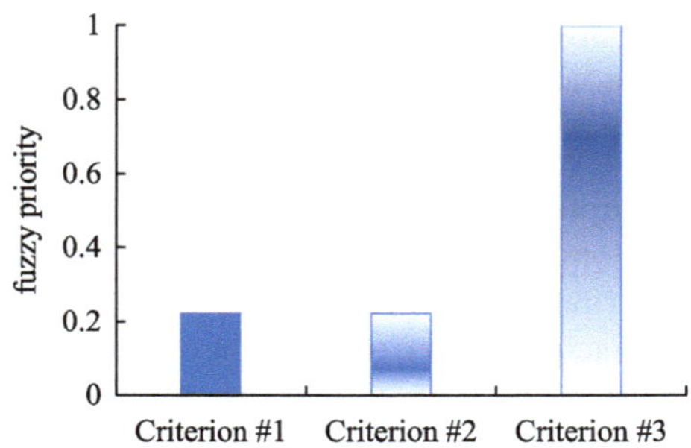

Fig. 3.7 Hanging gradient bar chart for the fuzzy priorities of criteria

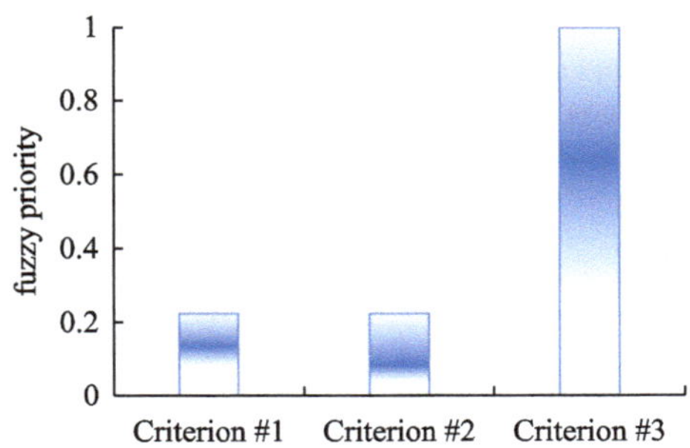

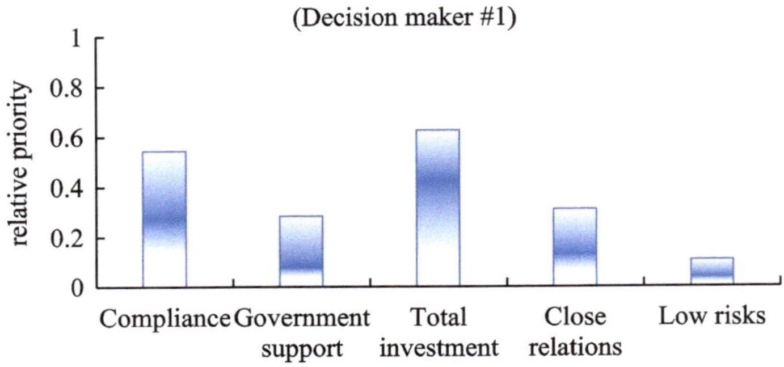

Fig. 3.8 Fuzzy priorities of criteria in Example 3.2

acceptable to the decision-maker. Equation (3.28) can confirm this again. Among all the derivation methods, the *SAD* when using SCDT will be the smallest.

Example 3.2 For the previous example, the fuzzy priorities of criteria are derived from the fuzzy judgment matrix using SCDT. The derivation results are illustrated using a hanging gradient bar chart (see Fig. 3.8). Common description technique is also applied to replace technical terms in this figure. The hanging gradient bar chart in Fig. 3.8 improved the understandability by 87% over a bar chart for the same purpose.

The performances in optimizing criteria can be placed next to the priorities of these criteria and ranked them in order of the priorities of criteria to illustrate how well a facility location performs [30], which results in a hanging group gradient bar chart, as shown in Fig. 3.9.

Example 3.3 In the previous example, the decision-makers consider eight facility locations that are located in different regions and have unequal advantages. The performances of these facility locations are evaluated based on the five criteria. By placing the performance in optimizing each criterion next to the priority of the criterion and ranking them in order of the priorities of criteria, a hanging group

Fig. 3.9 Hanging group gradient bar chart for the priorities and performances of criteria

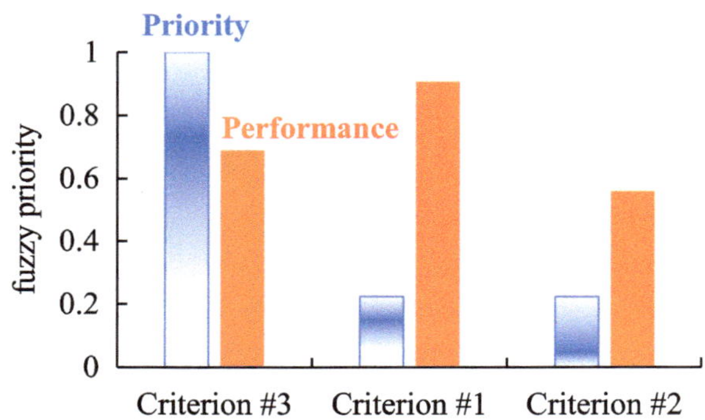

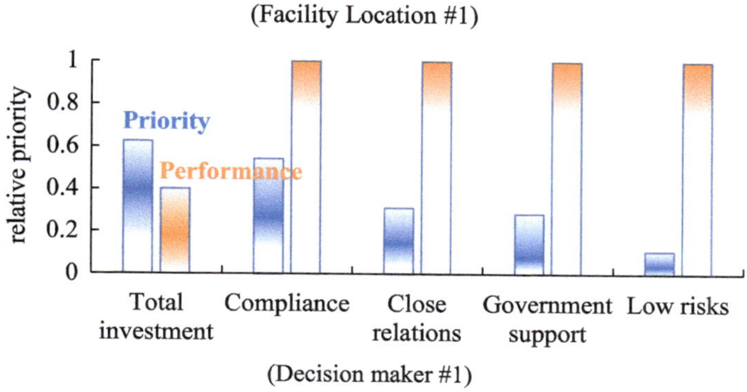

Fig. 3.10 Hanging group gradient bar chart for the priorities and performances of criteria of facility location #1

gradient bar chart is drawn to illustrate how well a facility location performs (see Fig. 3.10).

3.4.2 Interpreting the Overall Performance Evaluation of Facility Locations

Subsequently, GFTOPSIS method is proposed to evaluate and compare the overall performances of various facility locations:

$$\tilde{C}_q(m) = \frac{\xi \tilde{d}_q^-(m)}{(1-\xi)\tilde{d}_q^+(m) + \xi \tilde{d}_q^-(m)} \tag{3.33}$$

in which

$$\tilde{d}_q^+(m) = \sqrt{\sum_{i=1}^{n} (\tilde{\Lambda}_q^+(m)(-)\tilde{s}_{qi}(m))^\eta} \tag{3.34}$$

$$\tilde{d}_q^-(m) = \sqrt{\sum_{i=1}^{n} (\tilde{s}_{qi}(m)(-)\tilde{\Lambda}_q^-(m))^\eta} \tag{3.35}$$

where $\tilde{C}_q(m)$ is the fuzzy closeness (overall performance) of facility location q for decision-maker m; $\tilde{d}_q^+(m)$ and $\tilde{d}_q^-(m)$ indicate the fuzzy distances from facility location q to the ideal and anti-ideal points, respectively, for decision-maker m; $\{\tilde{\Lambda}_q^+(m)\}$ and $\{\tilde{\Lambda}_q^-(m)\}$ denote the fuzzy ideal and anti-ideal solutions, respectively; and $\tilde{s}_{qi}(m)$ is the fuzzy prioritized score of facility location q with respect to criterion i:

$$\tilde{s}_{qi}(m) = \tilde{w}_i(m)(\times)\tilde{\rho}_{qi} \tag{3.36}$$

where $\tilde{\rho}_{qi}$ is the normalized performance, which is independent of decision-makers.

The bidirectional scatterplot designed by Wang and Chen [31] cannot visualize the uncertainty of the overall performance of a facility location, as shown in Fig. 3.11. Nevertheless, the effectiveness of a bidirectional scatterplot can be assessed as follows. In Fig. 3.11, the scores of ideal and anti-ideal solutions are set to 10 and 0, respectively. Then, a viewer is asked to estimate the score of each facility location, based on which

$$\widetilde{SAD} = \sum_q \left|\hat{\tilde{S}}_q(m)(-)\tilde{C}_q(m)(10-0)+0\right| \tag{3.37}$$

where $\hat{\tilde{S}}_q(m)$ denotes the (fuzzy) score of facility location #q estimated by the viewer. To facilitate the calculation, $\widetilde{SAD}$ is defuzzified as

$$SAD = \sum_q \left(\left|\hat{S}_{q1}(m) - 10C_{q1}(m)\right| + 2\left|\hat{S}_{q2}(m) - 10C_{q2}(m)\right| + \left|\hat{S}_{q3}(m) - 10C_{q3}(m)\right|\right) \tag{3.38}$$

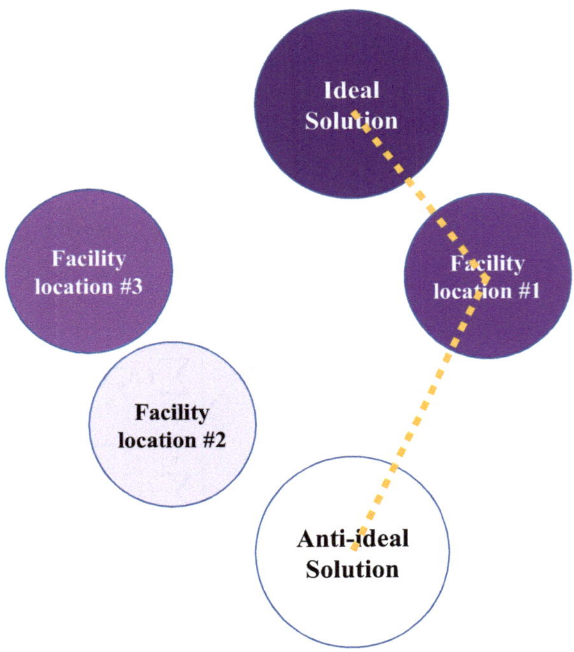

Fig. 3.11 Bidirectional scatterplot for the overall performances of facility locations

$$SAD = \sum_{q} \left(|\hat{S}_{q1}(m) - 10C_{q1}(m)| + 2 \cdot |\hat{S}_{q2}(m) - 10C_{q2}(m)| + |\hat{S}_{q3}(m) - 10C_{q3}(m)| \right) \tag{3.39}$$

For example, after viewing Fig. 3.11, a viewer estimated the scores of the three facility locations as $\hat{\tilde{S}}_1(m) = 7.5$, $\hat{\tilde{S}}_2(m) = 2$, and $\hat{\tilde{S}}_3(m) = 4.7$, while $\tilde{C}_1(m) = (0.577, 0.638, 0.698)$, $\tilde{C}_2(m) = (0.322, 0.348, 0.374)$, and $\tilde{C}_3(m) = (0.593, 0.637, 0.681)$, then

$$\begin{aligned} SAD =& |7.5-10 \cdot 0.577| + 2 \cdot |7.5-10 \cdot 0.637| + |7.5-10 \times 0.698| + |2-10 \cdot 0.322| \\ & + 2 \cdot |2-10 \cdot 0.348| + |2-10 \cdot 0.374| + |4.7-10 \cdot 0.593| \\ & + 2 \cdot |4.7-10 \cdot 0.637| + |4.7-10 \cdot 0.681| = 17.11 \end{aligned}$$

However, the uncertainty in the distance is not revealed, which can be fixed by devising a gradient bidirectional scatterplot, as shown in Fig. 3.12.

After viewing Fig. 3.12, the viewer estimated the following fuzzy scores: $\hat{\tilde{S}}_1(m) = (5.8, 7.5, 8)$, $\hat{\tilde{S}}_2(m) = (1.5, 2, 3)$, and $\hat{\tilde{S}}_3(m) = (4.1, 4.7, 5.3)$. Then,

$$\begin{aligned} SAD =& |5.8-10 \cdot 0.577| + 2 \cdot |7.5-10 \cdot 0.637| + |8-10 \cdot 0.698| + |1.5-10 \cdot 0.322| \\ & + 2 \cdot |2-10 \times 0.348| + |3-10 \cdot 0.374| + |4.1-10 \cdot 0.593| + 2 \cdot |4.7-10 \cdot 0.637| \\ & + |5.3-10 \cdot 0.681| = 15.41 \end{aligned}$$

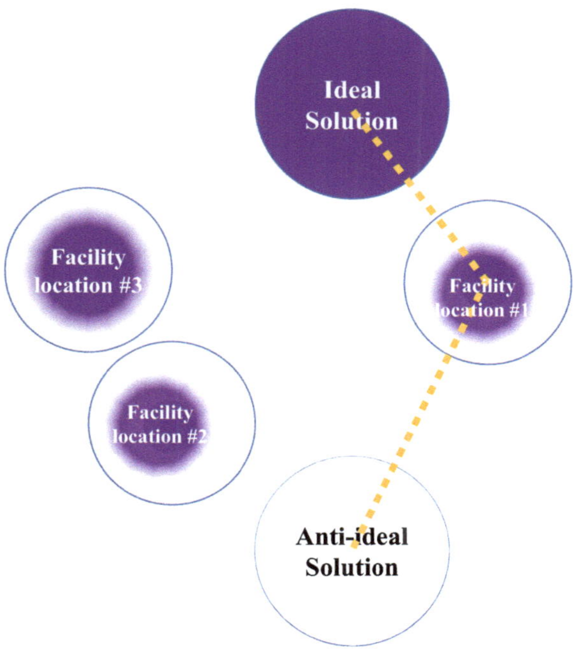

Fig. 3.12 Gradient bidirectional scatterplot for the overall performances of facility locations

which is better than the result after viewing Fig. 3.11.

Example 3.4 GFTOPSIS is applied to the previous example to evaluate and compare the overall performances of the eight facility locations. η and ξ are set to 2 and 0.5, respectively. The results are summarized in Fig. 3.13 using a gradient bidirectional scatterplot, showing the intensity of competition between these facility locations. The effectiveness of the gradient bidirectional scatterplot in enhancing the understandability of GFTOPSIS is assessed using Eq. (3.28). The performances of various methods for the same purpose were compared in Table 3.3.

To compare the suitabilities of various visualization XAI techniques, domain experts can be asked to evaluate the XAI performances of these visualization XAI technique in terms of specificity, recognizability, comprehensibility, ease of implementation, and speed of generation, using a Likert from 0 to 5.

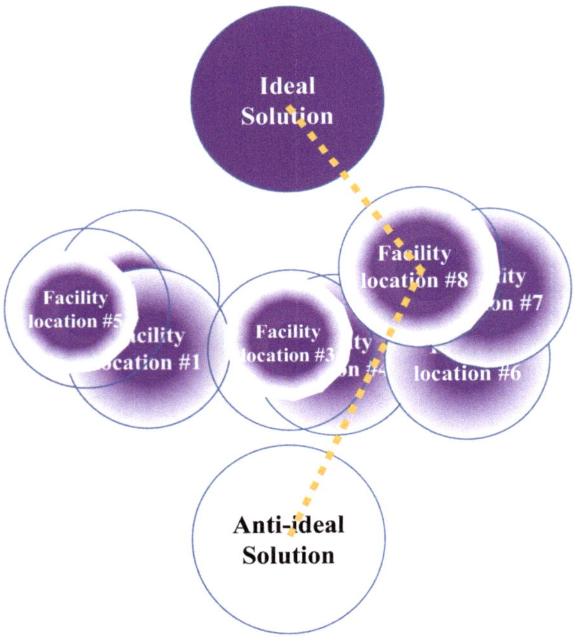

Fig. 3.13 Overall performances of the eight facility locations

Table 3.3 Comparison of the performances of various methods in enhancing the understandability

Figure	Bar chart	Bidirectional scatterplot	Gradient bidirectional scatterplot
SAD	12.34	16.25	5.14
Improvement %	–	–39%	58%
Explaining the reasoning mechanism	No	Yes	Yes

Example. 3.5 In the previous example, decision-makers evaluate the performances of various visualization XAI techniques in terms of specificity, recognizability, comprehensibility, ease of implementation, and speed of generation. Then, all decision-makers' evaluations are averaged. The evaluation results are summarized in Table 3.4.

Example 3.6 In the previous example, some experts are asked to ascertain whether the advantage of the gradient bidirectional scatterplot over the bidirectional scatterplot is significant. Then, a paired t test is conducted to examine the following hypotheses:

H_0: The explainability in terms of the confidence/trust scores achieved using the gradient bidirectional scatterplots would be the same as those achieved using the bidirectional scatterplots.

H_1: The explainability in terms of the confidence/trust scores achieved using the gradient bidirectional scatterplots would be higher than those achieved using the bidirectional scatterplots.

Table 3.5 summarizes the results of the paired t test. H_0 is rejected if $\alpha > p(T \leq t)$ in a one-tailed test. The gradient bidirectional scatterplots outperform the bidirectional scatterplots. The explainability of the gradient bidirectional scatterplots is significantly higher ($\alpha = 0.05$) than those achieved with the bidirectional scatterplots.

Example 3.7 For the previous example, a parametric analysis is conducted by gradually adjusting the relative priority of total investment with respect to compliance with local needs to see if the effectiveness of the hanging gradient bar chart changes. The results are summarized in Fig. 3.14. It is worth noting that the fuzzy judgment matrix becomes inconsistent when the relative priority exceeded 5. Obviously, the hanging gradient bar chart still remains highly effective, which supports its robustness.

Table 3.4 Performances of various visualization XAI techniques (by averaging all decision-makers' evaluations)

Criterion	Specificity	Recognizability	Comprehensibility	Ease of implementation	Speed of generation
Bidirectional scatterplot	4.26	2.14	3.90	1.78	0.90
Gradient bidirectional scatterplot	4.56	2.93	4.01	1.79	3.44

Table 3.5 Results of the paired t test

	Existing method (bidirectional scatterplots)	Proposed methodology (gradient bidirectional scatterplots)
Mean	3.429	3.831
Variation	0.175	0.206
Observations	23	23
Pearson correlation coefficient	0.348	NA
Degree of freedom	22	NA
t statistic	-3.867	NA
$p(T \leq t)$ one-tail	4.17E-04	NA
t Critical one-tail	1.717	NA
$p(T \leq t)$ two-tail	8.33E-04	NA
t critical two-tail	2.074	NA

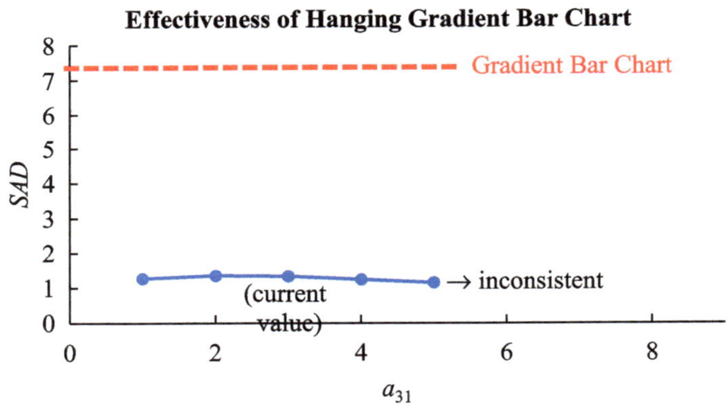

Fig. 3.14 Parametric analysis results

3.4.3 Interpreting the Evaluation Result Aggregation of All Experts

In most FGDM studies based on anterior aggregation [32], the opinions of decision-makers are aggregated at the outset and therefore analyzed only from a single perspective. To compensate for this, a posterior analysis can be conducted from the perspectives of all decision-makers and the overall performances of facility locations evaluated by them are aggregated using fuzzy intersection [33, 34]:

$$\tilde{C}_q = \widetilde{FI}\left(\left\{\tilde{C}_q(m)|m = 1 \sim M\right\}\right) \tag{3.40}$$

where

$$\mu_{\widetilde{FI}}(x) = \min_m(\mu_{\widetilde{C}_q(m)}(x)) \tag{3.41}$$

According to Lin and Chen [28], a traceable aggregation plot can be drawn to illustrate the aggregation process and results, as illustrated in Fig. 3.15. The aggregation result is also illustrated using a hanging gradient bar chart for traceable aggregation (see Fig. 3.16), which eliminates the need for a user to comprehend the shape of a membership function.

Equation (3.38) can also be used to evaluate the effectiveness of a hanging gradient bar chart for traceable aggregation similarly as

$$\widetilde{SAD} = \sum_m \left|\hat{\tilde{C}}_q(m)(-)\tilde{C}_q(m)\right| + \left|\hat{\tilde{C}}_q(-)\tilde{C}_q\right| \tag{3.42}$$

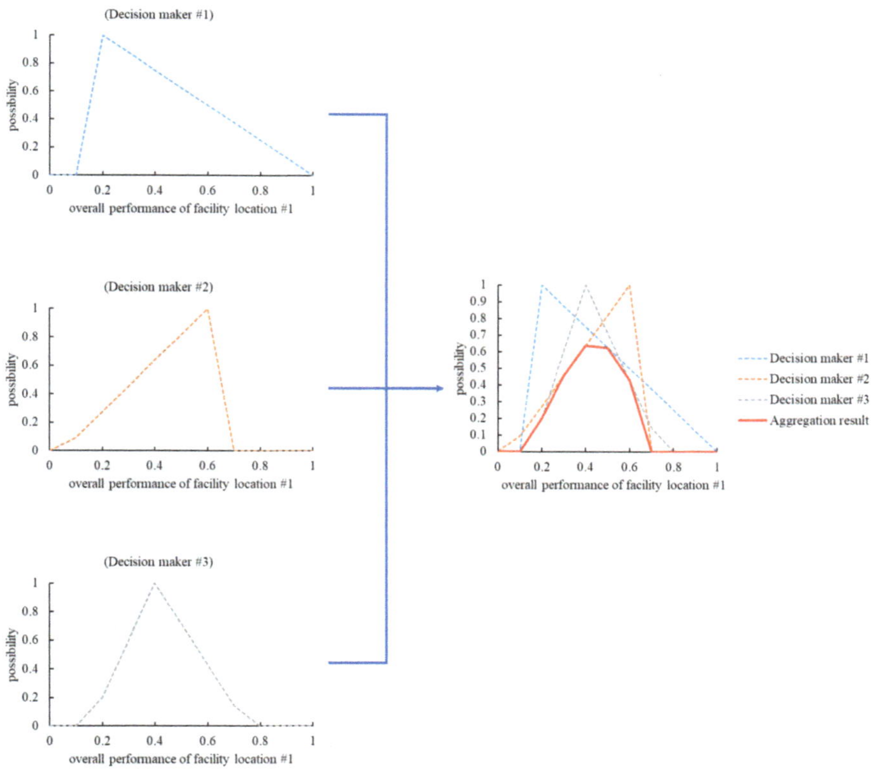

Fig. 3.15 Traceable aggregation plot for three decision-makers' evaluation results

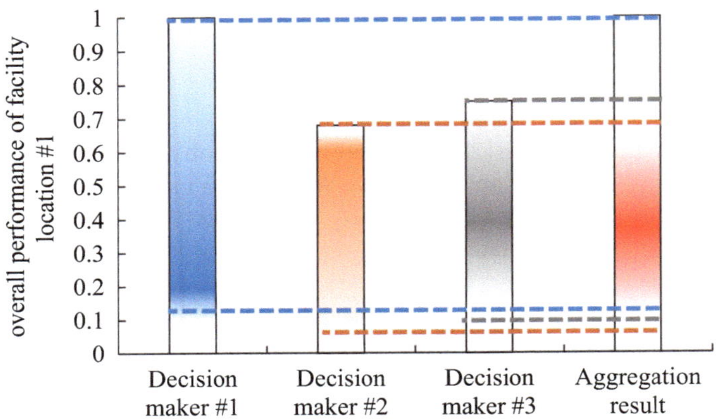

Fig. 3.16 Hanging gradient bar charts with traceable aggregation for three decision-makers' evaluation results

where $\hat{\tilde{C}}_q(m)$ and $\hat{\tilde{C}}_q$ denote the estimated values of $\tilde{C}_q(m)$ and $\tilde{C}_q$ after viewing Fig. 3.10. To facilitate the calculation, $\widetilde{SAD}$ is defuzzified as

$$SAD = \sum_m \left(\left| \hat{C}_{q1}(m) - C_{q1}(m) \right| + 2\left| \hat{C}_{q2}(m) - C_{q2}(m) \right| + \left| \hat{C}_{q3}(m) - C_{q3}(m) \right| \right) + \left| \hat{C}_{q1} - C_{q1} \right| + 2\left| \hat{C}_{q2} - C_{q2} \right| + \left| \hat{C}_{q3} - C_{q3} \right| \quad (3.43)$$

Example 3.8 In the previous example, the evaluation results by the three decision-makers are aggregated, which is illustrated using a hanging gradient bar charts for traceable aggregation in Fig. 3.17. The effectiveness of the hanging gradient bar chart for traceable aggregation was assessed using Eq. (3.31) as 2.65.

3.4.4 Evaluating Criterion Importance Using Shapley Additive Explanations (SHAP)

The SHAP value is defined as the weighted average of the marginal contributions of all possible coalitions $|\mathbf{F}|!$ as [35]

$$\varphi_m(f) = \sum_{\{\mathbf{S} \subseteq \mathbf{F}\}\{m\}} \frac{|\mathbf{S}|!(|\mathbf{F}| - |\mathbf{S}| - 1)!}{|\mathbf{F}|!} \circ \left[f\left(x_{\mathbf{S} \cup \{m\}}\right) - f(x_{\mathbf{S}}) \right] \quad (3.44)$$

where

$\varphi_m(f)$ is the weighted average Shapley value that feature m provides in the context of all coalitions that exclude feature m;

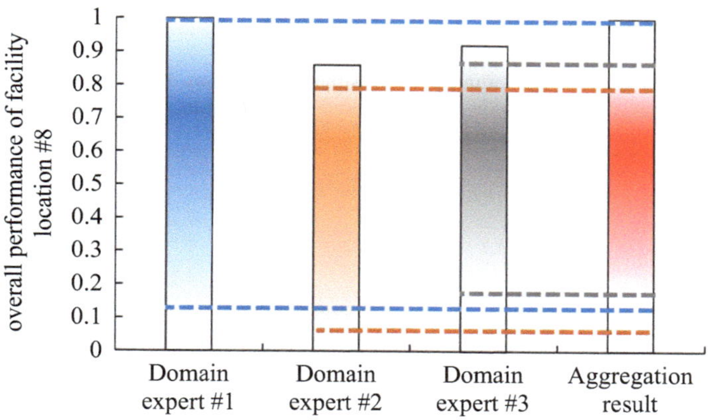

Fig. 3.17 Hanging gradient bar chart with traceable aggregation for the evaluation results of three decision-makers

F is the set of all features;

S is a subset (i.e., coalition) of **F**;

$(x_{\mathbf{S}\cup\{m\}})$ is the model prediction considering feature m, while $f(x_{\mathbf{S}})$ is the model prediction without considering feature m.

A special case is when $|\mathbf{S}| = |\mathbf{F}| - 1$, i.e., $\mathbf{S} = \mathbf{F}\backslash\{m\}$,

$$\varphi_m(f) = \sum_{\{\mathbf{S}\subseteq\mathbf{F}\}\{m\}} \frac{|\mathbf{S}|!0!}{|\mathbf{F}|!} \circ \left[f\left(x_{\mathbf{S}\cup\{m\}}\right) - f(x_{\mathbf{S}})\right] = \sum_{\{\mathbf{S}\subseteq\mathbf{F}\}\{m\}} \frac{\left[f\left(x_{\mathbf{S}\cup\{m\}}\right) - f(x_{\mathbf{S}})\right]}{|\mathbf{F}|!} \tag{3.45}$$

In this way, SHAP evaluates the importance of an input by fixing its value while randomizing other inputs, and then averaging the resulting differences in $f()$.

The importance of a criterion is different for different facility locations because decision-makers have different relative priorities for the same criterion. Therefore, if anterior aggregation is conducted, the importance of the criterion can be assessed. Otherwise, after posterior aggregation, the importance of the criterion cannot be determined and may vary depending on the facility location.

Example 3.9 For a facility location selection problem in a supply chain, five criteria are considered: compliance with local needs, local government support and subsidies, total investment required, close relations with third parties, and low political and war risks. In addition, five possible facility locations are evaluated. The evaluation results are summarized in Table 3.6.

The importance of each criterion for facility location #1 is evaluated in terms of a SHAP value. The results are summarized by a waterfall plot in Fig. 3.18.

For comparison, the evaluation results for facility location #4 are shown in Fig. 3.19. Although "compliance with local needs" is the most important criterion for

Table 3.6 Evaluation results

Facility location #	Compliance with local needs	Local government support and subsidies	Total investment required	Close relations with third parties	Low political and war risks	FTOPSIS (Defuzzified value)
1	2.16	3.67	1	4.77	1.97	0.01
2	4	5	4.4	5	5	0.99
3	1	2.33	5	3.32	1.86	0.41
4	4.64	5	4.4	1	5	0.99
5	5	5	3.7	3.66	1	0.70

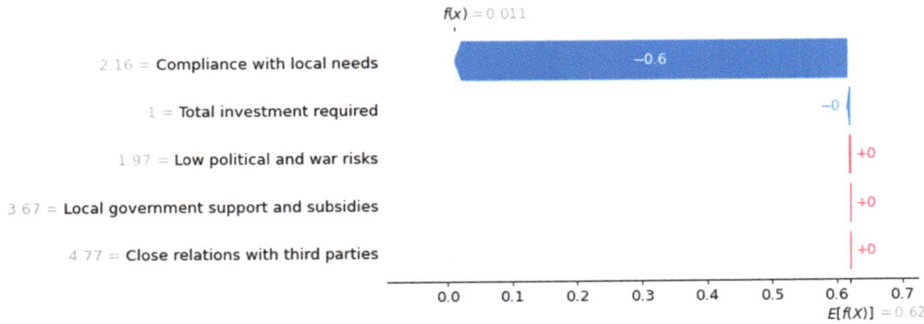

Fig. 3.18 SHAP values of all criteria for facility location #1

the two facility locations, its effects are different. For facility location #1, its effect is negative, namely, it decreases the overall performance. In contrast, for facility location #4, it has a positive effect by increasing the overall performance of this facility location.

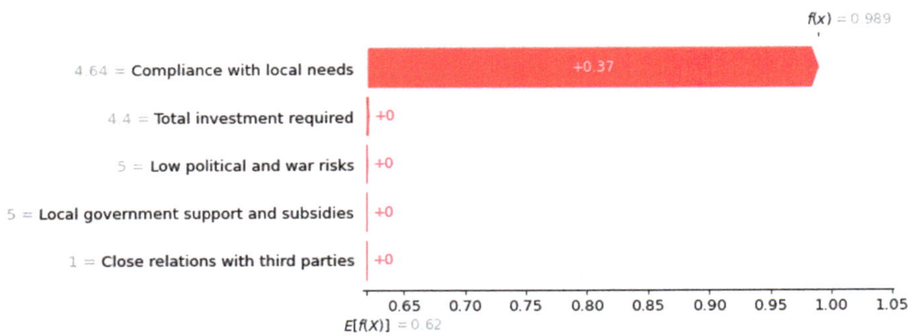

Fig. 3.19 SHAP values of all criteria for facility location #4

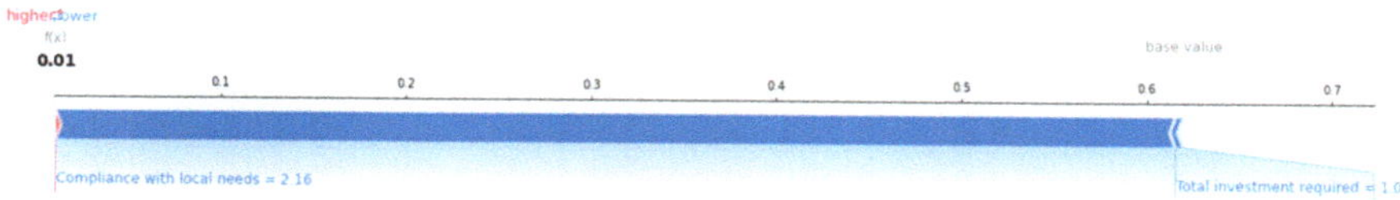

Fig. 3.20 Force plot for facility location #1

The evaluation results for facility location #1 are summarized using a force plot in Fig. 3.20, showing the incremental effect of each criterion in contributing to the overall performance of this facility location.

Waterfall plots and force plots are applicable for generating local explanations, while scatter plots and beeswarm plots can provide global explanations for all facility locations, as shown in Figs. 3.21 and 3.22. In Fig. 3.21, the effects of "compliance with local needs" for all facility locations are summarized, showing that when the performance in optimizing "compliance with local needs" is better (in red), the SHAP value becomes higher. Figure 3.22 shows that the importance of "compliance with local needs" for all facility locations is more uncertain.

However, criterion importance in terms of a SHAP value is evaluated by fixing the performance in optimizing the criterion while randomizing the performances in optimizing the others. However, the performance of a facility location in optimizing a criterion cannot be easily changed, causing difficulties in evaluating the SHAP value.

Local interpretable model-agnostic explanations (LIME) aims to avoid overfitting and increase transparency by using simpler models, such as linear regression (LR) or decision trees (DT), to fit the local feasible region around an example. An example is provided in Fig. 3.23, which presents a LIME summary plot. The left side of the figure shows the output (or predicted value) of the case relative to the output range.

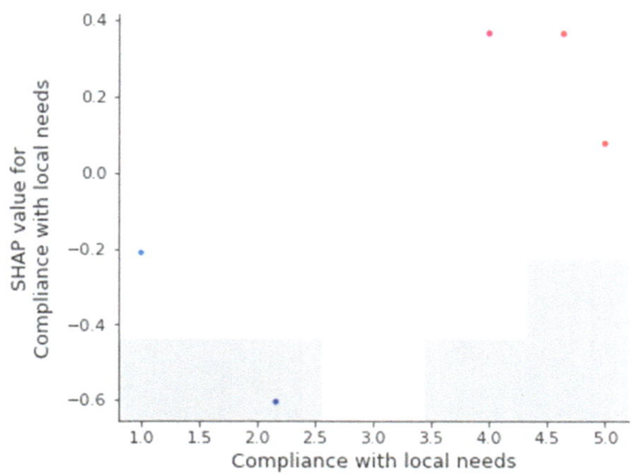

Fig. 3.21 Scatter plot

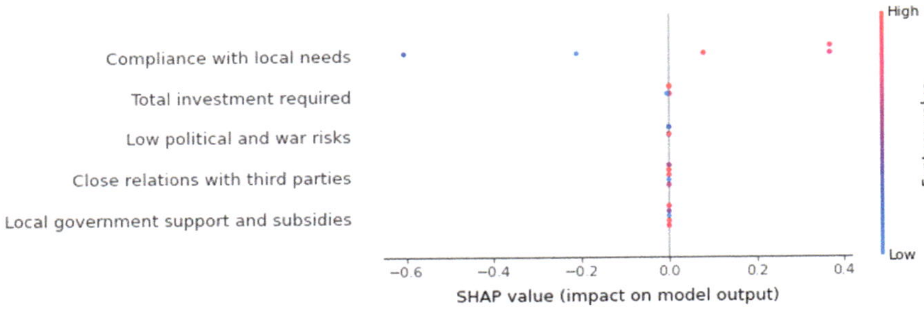

Fig. 3.22 Beeswarm plot

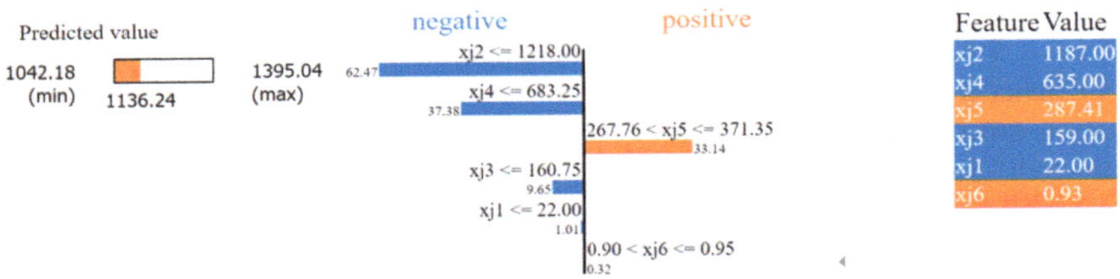

Fig. 3.23 LIME summary plot

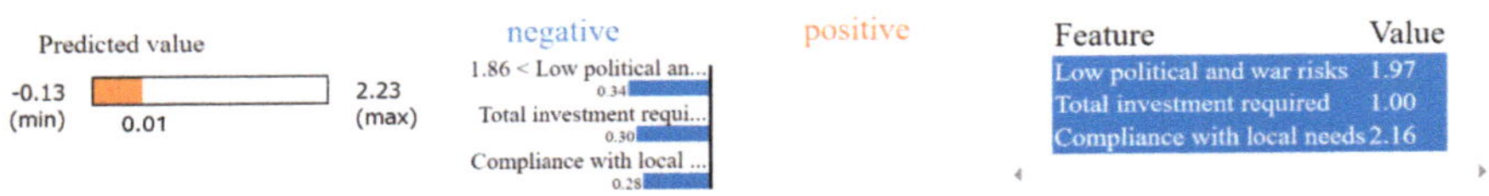

Fig. 3.24 LIME summary plot for facility location #1

Additionally, the ride side provides the value of each input. Based on a simpler model, the contribution of each input variable to the output is also evaluated. Similar to the waterfall model in SHAP, positive and negative contributions are represented by different colors.

Example 3.10 For the previous example, the LIME summary report for the first facility location is shown in Fig. 3.24.

References

1. O. Kilincci, S.A. Onal, Fuzzy AHP approach for supplier selection in a washing machine company. Expert Syst. Appl. **38**(8), 9656–9664 (2011)
2. C. Kahraman, D. Ruan, I. Doğan, Fuzzy group decision-making for facility location selection. Inf. Sci. **157**, 135–153 (2003)

3. T. C. T. Chen, Issues in managing localized semiconductor supply chains, in *Supply Chain Localization in the Semiconductor Industry: Rebuilding the Competitiveness and Sustainability of Semiconductor Manufacturers* (2025), pp. 45–71
4. T. Chen, Y.C. Wang, P.H. Jiang, A selectively calibrated derivation technique and generalized fuzzy TOPSIS for semiconductor supply chain localization assessment. Decis. Anal. J. **8**, 100275 (2023)
5. T.C.T. Chen, Localization of semiconductor supply chains—Driving forces and challenges, in *Supply Chain Localization in the Semiconductor Industry: Rebuilding the Competitiveness and Sustainability of Semiconductor Manufacturers* (2025), pp. 23–44
6. R. Abudu, E.A. Thompson, F.S. Dzawu, A. Roa-Henriquez, A meta-analysis assessment of adaptive and transformative approaches to supply chain resilience. Supply Chain Anal. **10**, 100124 (2025)
7. T.C.T. Chen, Rebuilding semiconductor manufacturing competitiveness and sustainability through supply chain localization, in *Supply Chain Localization in the Semiconductor Industry: Rebuilding the Competitiveness and Sustainability of Semiconductor Manufacturers* (2025), pp. 127–160
8. Z. Wu, F. Jia, Toward a theory of supply chain fields–understanding the institutional process of supply chain localization. J. Oper. Manag. **58**, 27–41 (2018)
9. R. Reza-Gharehbagh, A. Hafezalkotob, S. Asian, A. Makui, A.N. Zhang, Peer-to-peer financing choice of SME entrepreneurs in the re-emergence of supply chain localization. Int. Trans. Oper. Res. **27**(5), 2534–2558 (2020)
10. T.C.T. Chen, *Supply Chain Localization in the Semiconductor Industry* (Springer Nature, 2025)
11. TrendForce.com, [News] Potential U.S. tariffs on Taiwan semiconductors spark calls for stronger U.S.-Taiwan tech cooperation (2025). https://www.trendforce.com/news/2025/02/11/news-potential-u-s-tariffs-on-taiwan-semiconductors-spark-calls-for-stronger-u-s-taiwan-tech-cooperation/
12. S. Belgacem, K. Murakami, Mitigating the impact of COVID 19 and the Russian-Ukrainian war on the semiconductor industry strategy analysis of investing in localizing the semiconductor supply chain in Japan. Global Bus. J. **10**(1), 11–22 (2024)
13. TechNews, Semiconductor leaders cannot invest in the United States without the chemical industry Taiwan team (2020). https://technews.tw/2020/06/29/tsmc-big-business-opportunities-for-semiconductor-materials-in-taiwan/
14. Q. Merle, Chips supply chain: bifurcation and localization. CSS Anal. Secur. Policy **345**, 1–4 (2024)
15. J. An, H. Yin, EU policies and prospects for achieving 'Chip autonomy' amid global localization trend in chip supply chain. Innovat. Eur. J. Soc. Sci. Res. 1–22 (2025)
16. S. Luthra, K. Govindan, D. Kannan, S.K. Mangla, C.P. Garg, An integrated framework for sustainable supplier selection and evaluation in supply chains. J. Clean. Prod. **140**, 1686–1698 (2017)
17. S. Mohseni, K. Baghizadeh, J. Pahl, Evaluating barriers and drivers to sustainable food supply chains. Math. Probl. Eng. **2022**(1), 4486132 (2022)
18. S. Kheybari, F.M. Rezaie, Selection of biogas, solar, and wind power plants' locations: an MCDA approach. J. Supply Chain Manag. Sci. **1**(1–2), 45–71 (2020)
19. T.C.T. Chen, Introduction to fuzzy group decision-making, in *Advances in Fuzzy Group Decision Making* (2021), pp. 1–9
20. T.C.T. Chen, Y.C. Wang, H.C. Wu, An explainable fuzzy collaborative intelligence approach for prioritizing possible measures to tackle challenges in semiconductor manufacturing supply chain localization. Oper. Res. Int. J. **26**(1), 7 (2026)
21. M.C. Chiu, T.C.T. Chen, Prioritizing measures to mitigate the impact of cultural differences on semiconductor supply chain localization: a neutrosophic collaborative intelligence approach. Cogn. Comput. **17**(4), 125 (2025)
22. A. Nafei, S.P. Azizi, S.A. Edalatpanah, C.Y. Huang, Smart TOPSIS: a neural network-driven TOPSIS with neutrosophic triplets for green supplier selection in sustainable manufacturing. Expert Syst. Appl. **255**, 124744 (2024)

23. Y.C. Wang, T.C.T. Chen, C.W. Lin, Fuzzy reinforcement learning approach for supporting ubiquitous manufacturing facility selection in drone manufacturing. Int. J. Adv. Manuf. Technol. **141**(3), 1893–1903 (2025)
24. T.C.T. Chen, C.W. Lin, An FGM decomposition-based fuzzy MCDM method for selecting smart technology applications to support mobile health care during and after the COVID-19 pandemic. Appl. Soft Comput. **121**, 108758 (2022)
25. F. Ahmed, K. Kilic, Fuzzy analytic hierarchy process: a performance analysis of various algorithms. Fuzzy Sets Syst. **362**, 110–128 (2019)
26. T.C.T. Chen, Y.C. Wang, M.C. Chiu, An efficient approximating alpha-cut operations approach for deriving fuzzy priorities in fuzzy multi-criterion decision-making. Appl. Soft Comput. **139**, 110238 (2023)
27. G. Mugurusi, P.N. Oluka, Towards explainable artificial intelligence (xai) in supply chain management: A typology and research agenda, in *IFIP International Conference on Advances in Production Management Systems* (2021), pp. 32–38
28. Y.C. Lin, T.C.T. Chen, Type-II fuzzy approach with explainable artificial intelligence for nature-based leisure travel destination selection amid the COVID-19 pandemic. Digit. Health **8**, 20552076221106320 (2022)
29. T.C.T. Chen, Explainable artificial intelligence (XAI) in manufacturing, in *Explainable Artificial Intelligence (XAI) in Manufacturing: Methodology, Tools, and Applications* (2023), pp. 1–11
30. S.Y. Chi, L.H. Chien, Why defuzzification matters: an empirical study of fresh fruit supply chain management. Eur. J. Oper. Res. **311**(2), 648–659 (2023)
31. Y.C. Wang, T. Chen, New XAI tools for selecting suitable 3D printing facilities in ubiquitous manufacturing. Complex Intell. Syst. **9**, 6813–6829 (2023)
32. T.C.T. Chen, Qualitative and quantitative analysis of semiconductor supply chain localization, in *Supply Chain Localization in the Semiconductor Industry: Rebuilding the Competitiveness and Sustainability of Semiconductor Manufacturers* (2025), pp. 73–125
33. Y.C. Wang, T.C.T. Chen, A partial-consensus posterior-aggregation FAHP method—supplier selection problem as an example. Mathematics **7**(2), 179 (2019)
34. T.C.T. Chen, Y.J. Lee, Smart technologies for fall detection and prevention, in *Smart and Healthy Walking: Toward Better Health and Life in Smart Cities* (2024), pp. 73–98
35. A.M. Salih, Z. Raisi-Estabragh, I.B. Galazzo, P. Radeva, S.E. Petersen, K. Lekadir, G. Menegaz, A perspective on explainable artificial intelligence methods: SHAP and LIME. Adv. Intell. Syst. **7**(1), 2400304 (2025)

Chapter 4
XAI Applications to Manage Supply Chain Risks and Resilience

4.1 Supply Chain Risks and Resilience

Global supply chains are facing various risks, such as natural risks (e.g., earthquakes, floods, fires, pandemics, and tsunamis) or man-made risks (e.g., terrorist attacks, accidents, supplier adulterations, wars, and cyber-attacks) [1]. Recently, there is a trend in the semiconductor industry—semiconductor supply chain localization [2, 3], in which semiconductor manufacturing capacity originally owned by wafer foundries is to be transferred back to the home countries of chip makers, which is owing to U.S.–China trade war, geopolitics, the coronavirus disease 2019 (COVID-19) pandemic, Ukrainian-Russian war, environmental awareness, and the need for transparency in semiconductor supply chains, which involve pandemic risks, war risks, and political risks. Facing such risks, the participants of supply chains are taking various actions to mitigate the impact of these risks, so as to maintain their competitiveness and pursue long-term sustainability [3].

There are various definitions of supply chain resilience [4], which can be defined as the adaptive capability of a supply chain to prepare for unexpected events, respond to disruptions, and recover from them by maintaining continuous operations at the desired level of connectedness and control over structure and functions. The ability of supply chain participants to cope with various risks clearly helps to improve the resilience of the supply chain.

4.2 Artificial Intelligence (AI) Applications in Managing Supply Chain Risks and Resilience

Various artificial intelligence (AI) technologies have been applied in managing supply chain risks and resilience. Especially, there are numerous reports in the literature on the applications of fuzzy inference systems (FIS) [5–13]. FISs have

T.-C. T. Chen, *Explainable Artificial Intelligence in Supply Chain Management*,
SpringerBriefs in Applied Sciences and Technology,
https://doi.org/10.1007/978-3-032-22564-1_4

been applied to support various functionalities of supply chain management (SCM), including performance evaluation [5, 7–10], risk and resilience management [6], supplier selection [13], demand prediction [11, 12], planning [11, 13], etc.

Some relevant references are reviewed as follows. Pourjavad and Shahin [5] constructed a Mamdani FIS to evaluate the green SCM performances of four industrial and household paint manufacturers in Iran. This approach was divided into two phases. The first phase consisted of four FISs that evaluate the performances of green design, green procurement, green manufacturing, and reverse logistics, respectively. Each FIS considered three to four inputs that represented the qualitative performance in each aspect, which had considerable uncertainty, justifying the adoption of fuzzy logic. Subsequently, the outputs from the four FISs were fed to another FIS composed of 625 fuzzy inference rules (FIRs) to evaluate the overall green SCM performance. In addition, five defuzzification methods, center of gravity (or area) (COG), bisector of area (BOA), mean (or middle) of maximum (MOM), smallest (or first) of maximum (SOM), and largest (or last) of maximum (LOM), were applied for comparison. The construction of FIRs in their study was based on expert knowledge. Such practices are common in this field, but it is difficult to validate the correctness of the FIRs.

Behret et al. [6] constructed four Mamdani FISs for supply chain risk management, the input variables of which were supply chain risks classified into categories such as transportation/distribution, manufacturing, order cycle, warehousing, and procurement. Outputs from the FISs included not on time deliveries, incompleteness, incorrectness, and damages and effects. Subsequently, the overall risk of the target supply chain was derived by averaging the four outputs.

Pabarja et al. [7] used a Mamdani FIS to evaluate the lean, agile, resilient, and green (LARG) characteristics of a hospital medical equipment supply chain. They first used four additional FISs to evaluate the four aspects of LARG. For example, when evaluating the lean characteristic, three inputs were considered: cost reduction, time reduction, and just-in-time (JIT). The outputs of the four FISs were used as inputs to evaluate LARG, with value ranging from 0 to 1. The FIS for evaluating LARG contained 625 FIRs. They also used the response surface method (RSM) to visualize the effect of each pair of inputs on the output and their interaction.

Ganga and Carpinetti [8] developed a Mamdani FIS to evaluate supply chain performance based on performance metrics at levels 1 and 2 of the supply council operations reference (SCOR) model [9], which comprised FIRs such as.

> IF Days sales outstanding is High AND Inventory days of supply is High AND Days payable outstanding is High THEN Net profit is Low.

In this way, the performances in supply chain reliability, supply chain flexibility, supply chain responsiveness, cost management, and asset management were evaluated. There were 25 rules just for asset management. They also applied two defuzzification methods, MOM and COA, for comparison. Clearly, most researchers were unsure how to choose the most suitable defuzzification method.

For evaluating the performance of sustainable SCM for three pipe and fitting companies in Iran, Efendigil and Önüt [10] proposed a hybrid approach. First, the fuzzy decision-making trial and evaluation laboratory (DEMATEL) method was applied to exclude criteria with less influence in advance. Only the top five factors were used. Subsequently, a Mamdani FIS system based on the five factors was built to evaluate the performance of sustainable SCM for each pipe and fitting company, in which various defuzzification techniques were also applied for comparison.

To address the irregularity and uncertainty of demand in supply chain planning, Jung and Jeong [11] proposed a hybrid fuzzy approach. They first constructed a fuzzy linear programming (FLP) model for supply chain planning, in which demand was defined as a fuzzy parameter. Subsequently, they constructed an FIS that transformed fuzzy demand into crisp demand by considering the judgments of internal and external experts on future demand, thereby reflecting the expected demand irregularity.

Momenitabar et al. [12] constructed a demand forecasting model for a supply chain in the Iranian tire industry based on a Mamdani FIS. The input variables of the FIS were selected by 30 experts using the fuzzy Delphi method. Finally, five input variables were determined as product price, product quality, environmental impact, advertising investment, and product supply. The output of the FIS, fuzzy demand forecast, was input into a fuzzy goal programming model to plan the supply chain to minimize total cost.

Amindoust and Saghafinia [13] proposed an FIS ensemble approach for supplier selection in the textile and apparel industry. First, they constructed four FISs to evaluate the economic, environmental, environmental, and social compatibilities of each supplier. Then, they used two additional FISs to aggregate the outputs of the first four FISs to evaluate the economic and environmental-social compatibilities of these suppliers. Finally, they used the seventh FIS to combine the outputs from the two FISs, thereby determining the priorities of the suppliers.

A comparison of SCM methods based on FIS applications is presented in Table 4.1. These studies used fuzzy sets to account for qualitative variables, difficulties, or uncertainties in collecting or evaluating variables, subjective variables, etc., as summarized in Table 4.2. Furthermore, Mamdani-type FISs are clearly the most prevalent FISs in SCM. These systems retain fuzziness (uncertainty) until the final step, providing more information than clear values by generating fuzzy-valued outputs, i.e., being more informative. Managers are also free to choose defuzzification methods that suit their needs [14], offering significant flexibility.

In addition, much research in this field employed a hybrid approach [10–12], equipping another model (such as a mathematical programming model) that received the output of the FIS or provided the necessary input to the FIS.

Table 4.1 Comparison of SCM methods based on FIS applications

References	Topic	FIS type	Number of FIRs	Type of fuzzy variables	Defuzzification method
Pourjavad and Shahin [5]	Green SCM performance evaluation	Mamdani	625 (for phase two)	Trapezoidal fuzzy number (TrFN)	COG, BOA, MOM, SOM, LOM
Behret et al. [6]	Supply chain risk management	Mamdani	30	Triangular fuzzy number (TFN)	COG
Pabarja et al. [7]	Hospital medical equipment supply chain LARG evaluation	Mamdani	625	Gaussian fuzzy number	COG
Ganga and Carpinetti [8]	Supply chain performance prediction	Mamdani	25 (for asset management)	TFN	MOM, COG
Efendigil and Önüt [10]	Sustainable SCM performance evaluation	Mamdani	625	TrFN	COG, BOA, MOM, SOM, LOM
Jung and Jeong [11]	Supply chain planning	Mamdani	9	TrFN	COG
Momenitabar et al. [12]	Supply chain demand forecasting	Mamdani	16	TrFN	COG
Amindoust and Saghafinia [13]	Supply chain supplier selection	Mamdani	175	TFN	COG, BOA, MOM, SOM, LOM

4.3 FIS Applications in Supply Chain Risks and Resilience Management

Take the Mamdani FIS established by Ganga and Carpinetti [8] as an example. In this FIR, "days sales outstanding", "inventory days of supply", and "days payable outstanding" are fuzzy variables which values are given in linguistic values such as "low", "medium", and "high", forming a term set. Triangular fuzzy numbers (TFNs) are widely used to represent these linguistic values, as illustrated in Fig. 4.1. The membership function of a TFN $\tilde{A} = (A_1, A_2, A_3)$ is

$$\mu_{\tilde{A}}(x) = \begin{cases} \frac{x-A_1}{A_2-A_1} & if \quad A_1 \leq x < A_2 \\ \frac{x-A_3}{A_2-A_3} & if \quad A_2 \leq x < A_3 \\ 0 & otherwise \end{cases} \tag{4.1}$$

Table 4.2 Reasons for adopting fuzzy sets in SCM

Reason	Example
Qualitative variables	• Employee well-being and job satisfaction • Customer satisfaction • Etc.
Difficulties or uncertainties in collecting or evaluating variables	• Environmental risk • Health status and risks of employee • Health and safety practices • Life-cycle assessment performed • Etc.
Subjective variables	• Importance of criteria • Design for the environment • Customer satisfaction regarding service quality • Employees' and customers' awareness of environmental issues • Supplier service quality level • Etc.

For example, in Fig. 4.1, "days payable outstanding" is low can be represented with a TFN (30, 30, 51) [8]. The term set shown in Fig. 4.1 contains three linguistic values (and the corresponding TFNs).

Linguistic values at the extremes such as "low" and "high" can be better represented with trapezoidal fuzzy numbers (TrFNs) $\tilde{A} = (A_1, A_2, A_3, A_4)$, the membership of which is

$$\mu_{\tilde{A}}(x) = \begin{cases} \frac{x-A_1}{A_2-A_1} & if \quad A_1 \le x < A_2 \\ 1 & if \quad A_2 \le x < A_3 \\ \frac{x-A_4}{A_3-A_4} & if \quad A_3 \le x < A_4 \\ 0 & \quad otherwise \end{cases} \tag{4.2}$$

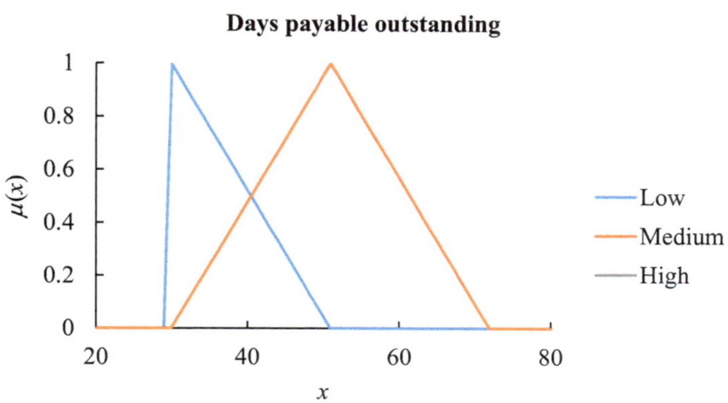

Fig. 4.1 Linguistic values given in TFNs

as illustrated in Fig. 4.2. For example, "days payable outstanding" is low is represented with a TrFN (0, 0, 30, 51) instead.

The definitions of these FIRs are based on the causal relationships proposed by SCOR [9]. The process for establishing an FIS is shown in Fig. 4.3.

In Mamdani FISs, the membership (or satisfaction level) of an example in each condition is used to truncate the membership function of the output in each FIR, as illustrated in Fig. 4.4, in which the x value of an example is 1.76, the membership

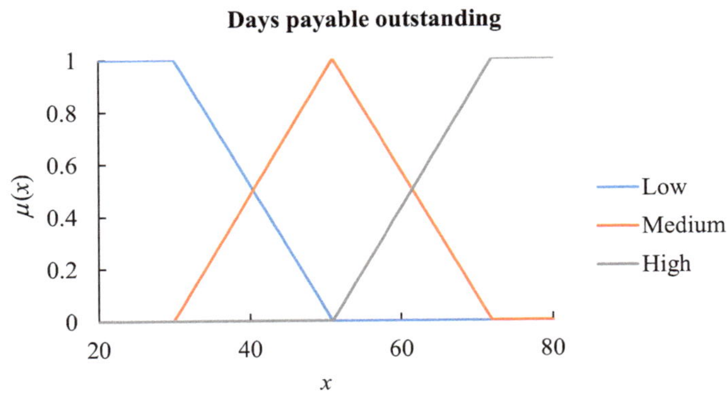

Fig. 4.2 Representing extreme values with TrFNs

Fig. 4.3 Process for establishing an FIS

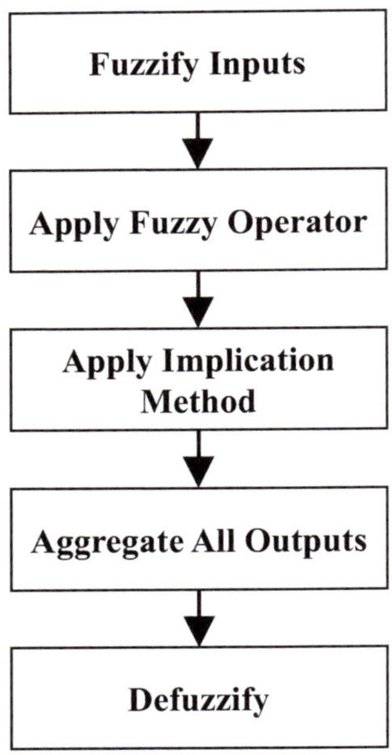

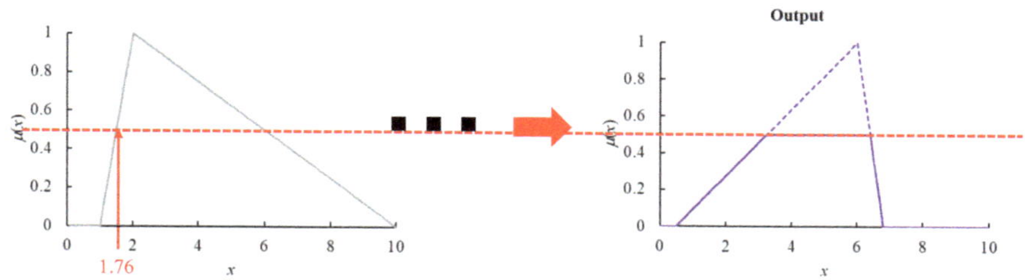

Fig. 4.4 Truncating the membership function of the output

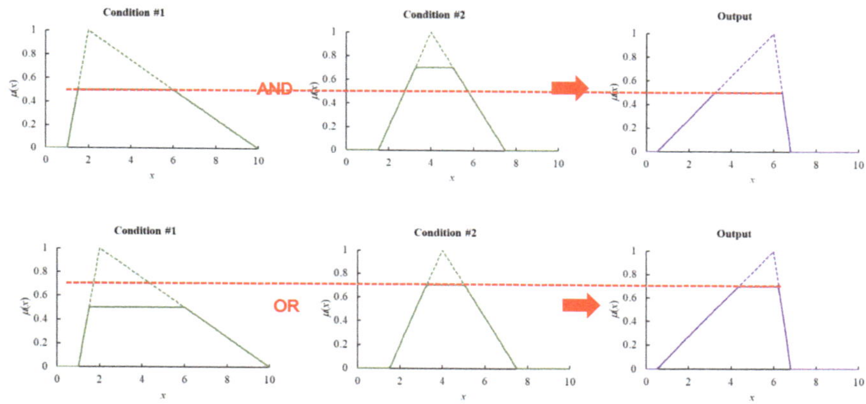

Fig. 4.5 Results using various logic operators

of which is 0.5. Therefore, the membership function of the output with membership greater than 0.5 is truncated and will not be used in aggregation.

Then, if the logic operator is AND, the minimum of the memberships of all conditions is used to truncate the membership function of the output variable. If the logic operator is OR, their maximum is used instead, as shown in Fig. 4.5. Another possible aggregator is product/multiplication, in which the memberships of an example in all conditions are multiplied to scale the output.

As a result, an FLR can be represented as $\wedge_m^k \tilde{C}_{km}(x_{jm}) \rightarrow \tilde{O}_k$, where x_{jm} is the mth attribute of example j; $\tilde{C}_{km}$ is the condition of attribute m in the kth FIR; and $\wedge_m^k$ is the logical operator for $\tilde{C}_{km}$.

Subsequently, the outputs from all FIRs are aggregated using fuzzy union (FU), as illustrated in Fig. 4.6:

$$\mu_{\tilde{A}}(x) = \mu_{\widetilde{FU}(\{\tilde{O}_k\})}(x) = \max_k \mu_{\tilde{O}_k}(x) \tag{4.3}$$

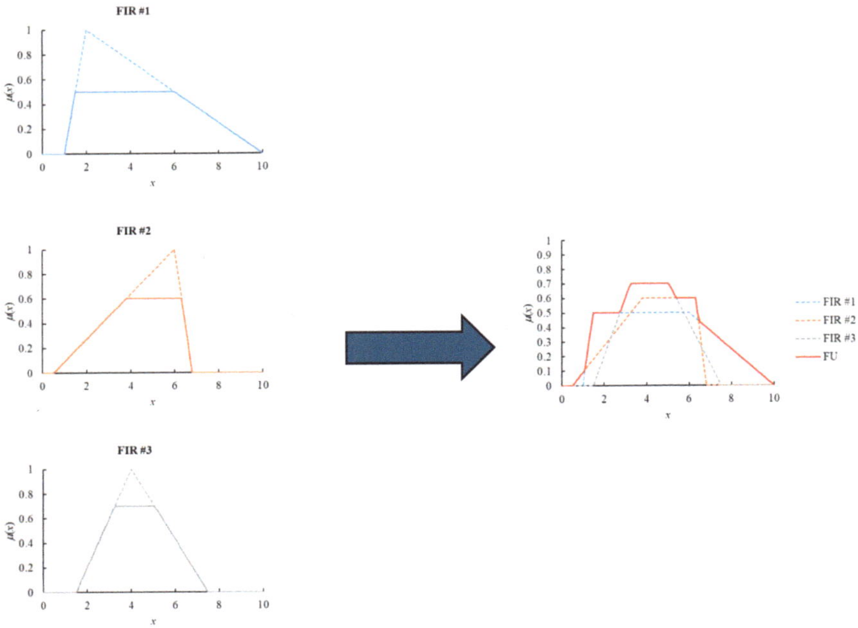

Fig. 4.6 Aggregation mechanism in Mamdani FISs

Then, the COG of the aggregation result is derived as the representative value, i.e., COG defuzzification [15], as shown in Fig. 4.7:

$$COG(\tilde{A}) = \frac{\int_{-\infty}^{\infty} x \cdot \mu_{\tilde{A}}(x)dx}{\int_{-\infty}^{\infty} \mu_{\tilde{A}}(x)dx} \quad (4.4)$$

where $\tilde{A}$ is the aggregation result. Other common defuzzification methods refer to Table 4.3. In contrast, in a Sugeno-type FIS, the aggregation is given by

$$A = \frac{\sum_k w_k O_k}{\sum_k w_k} \quad (4.5)$$

where w_k is firing strength of rule k:

$$w_k = \wedge_m^k \mu_{\tilde{C}_{km}}(x_{jm}) \quad (4.6)$$

Theorem 4.1 *When $\tilde{A}$ is a TFN $COG(\tilde{A}) = \frac{a_1+a_2+a_3}{3}$.*

Theorem 4.2 *When $\tilde{A}$ is a TrFN $COG(\tilde{A}) = \frac{1}{3}\left(a_1 + a_2 + a_3 + a_4 + \frac{a_1a_2 - a_3a_4}{a_3+a_4-a_1-a_2}\right)$* [17].

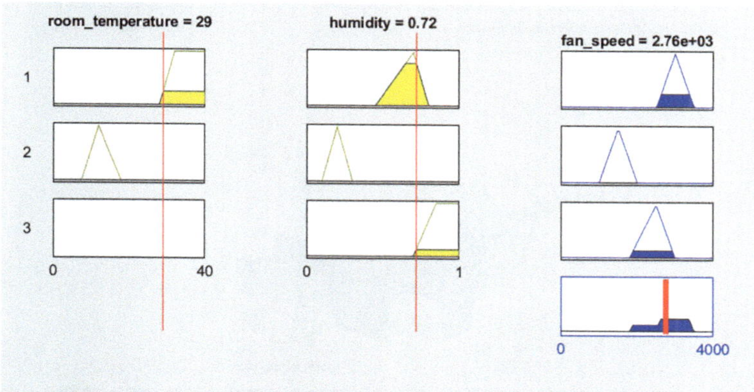

Fig. 4.7 Defuzzifying the aggregation result using COG

Table 4.3 Common defuzzification methods [16]

Defuzzification method	Formula
COG	$COG(\tilde{A}) = \frac{\int_{-\infty}^{\infty} x \cdot \mu_{\tilde{A}}(x)dx}{\int_{-\infty}^{\infty} \mu_{\tilde{A}}(x)dx}$
FOM	$FOM(\tilde{A}) = \min \text{core}(\tilde{A})$
LOM	$LOM(\tilde{A}) = \max \text{core}(\tilde{A})$
MOM	$LOM(\tilde{A}) = \frac{\min \text{core}(\tilde{A}) + \max \text{core}(\tilde{A})}{2}$
Basic defuzzification distributions (BADD)	$BADD(\tilde{A}) = \frac{\int_{-\infty}^{\infty} x \cdot \mu_{\tilde{A}}^{\gamma}(x)dx}{\int_{-\infty}^{\infty} \mu_{\tilde{A}}^{\gamma}(x)dx}$ $\gamma \in [0, \infty)$

As a result, an FLS can be represented as $\{FLR_k | FLR_k : \wedge_m^k \tilde{C}_{km}(x_{jm}) \rightarrow \tilde{O}_k; k = 1 \sim K\}$.

4.4 Traditional Methods for Interpreting FIS Applications

For an FIS application, the effects of the inputs, as well as their interaction, can be visualized using response surface method (RSM). An example is given in Fig. 4.8.

Example 4.1 To evaluate the responsiveness of a supply chain, three input variables, delivery commit date, inventory days of supply, and delivery cycle time, are considered. Input and output variables are mapped into TFNs shown in Table 4.4, using data from Ganga and Carpinetti [8]. Three FIRs are established as.

IF Delivery commit date is High AND Delivery cycle time is High THEN Responsiveness is Low.

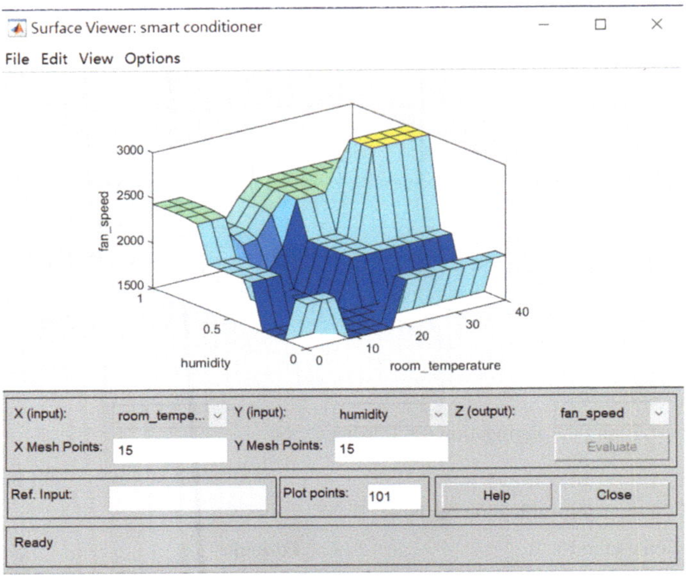

Fig. 4.8 RSM application [18]

Table. 4.4 Variables for evaluating the responsiveness of a supply chain

Variable	Low	Medium	High
Delivery commit date	(85%, 85%, 91.5%)	(85%, 91.5%, 98%)	(91.5%, 98%, 98%)
Inventory days of supply	(27.6, 27.6, 53.8) days	(27.6, 51.9, 80) days	(53.8, 80, 80) days
Delivery cycle time	(1, 1, 4) days	(1, 4, 7) days	(4, 7, 7) days
Responsiveness	(26.84%, 26.84%, 53.47%)	(26.84%, 53.47%, 99%)	(53.47%, 99%, 99%)

IF Delivery commit date is Low AND Inventory days of supply is Medium AND Delivery cycle time is Medium THEN Responsiveness is Medium.

IF Inventory days of supply is Low AND Delivery cycle time is Low THEN Responsiveness is High.

A Mamdani FIS composed of these FIRs is established using the built-in graphical user interface (GUI) of MATLAB, as shown in Fig. 4.9, in which the responsiveness of a supply chain with delivery commit date = 0.88, inventory days of supply = 39 days, and delivery cycle time = 6.3 days is evaluated.

The effects of each two inputs, as well as their interaction, are visualized using RSM in Fig. 4.10.

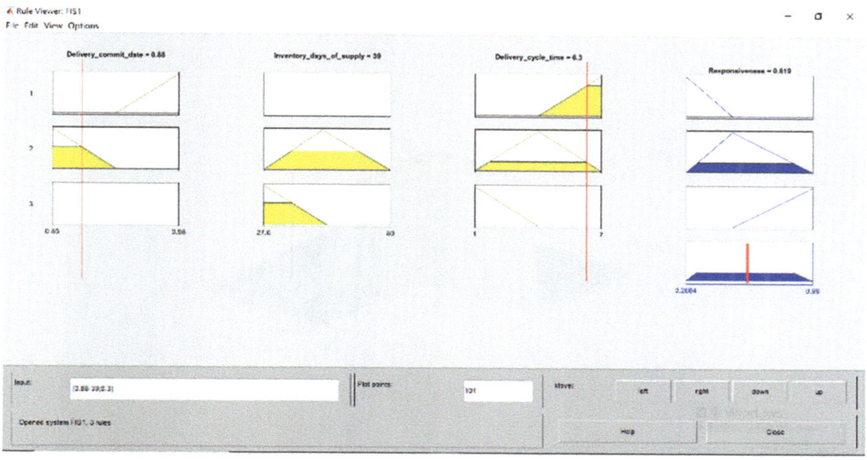

Fig. 4.9 Mamdani FIS for evaluating supply chain responsiveness

4.5 Explainable Artificial Intelligence (XAI) Methods for Interpreting FIS Application in SCM

There are several types of explainable artificial intelligence (XAI) methods: pre-model XAI methods, intrinsic XAI methods, and post hoc XAI methods [19]. In addition, XAI methods can also be divided into several categories such as visualization XAI methods [20], attribute importance evaluation XAI methods, and relationship approximation XAI methods [21]. Among visualization XAI methods, partial dependence plots, Shapley analysis (SHAP) plots (force plots, waterfall plots, scatter plots, and beeswarm plots), and local interpretable model-agnostic explanations (LIME) summary plots are among the most widely used techniques.

4.5.1 Partial Dependence Plot

A partial dependence plot (PDP) tells us about the interaction between the target response and an input feature of interest. The partial dependence (PD) of input variable m can be derived as

$$PD(x_m = \xi) = E\big(COG(\tilde{A})(x_m = \xi)\big) \tag{4.7}$$

Example 4.2 For Example 4.1, a one-way PDP is drawn for each input variable. The required MATLAB code for deriving the PD of each input variable is provided in Fig. 4.11. The results are summarized in Fig. 4.12.

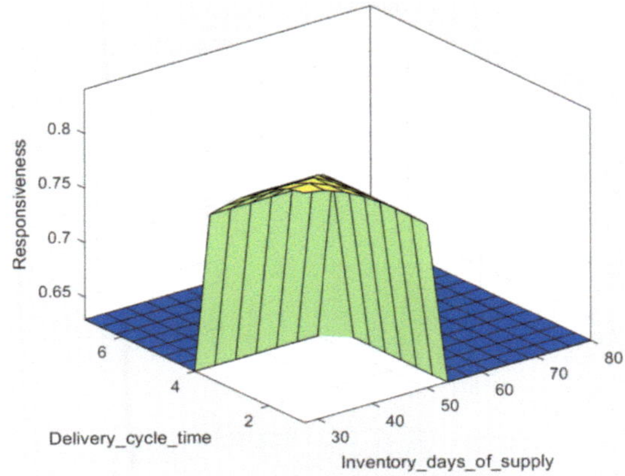

(Effects of "delivery cycle time" and "inventory days of supply" on responsiveness)

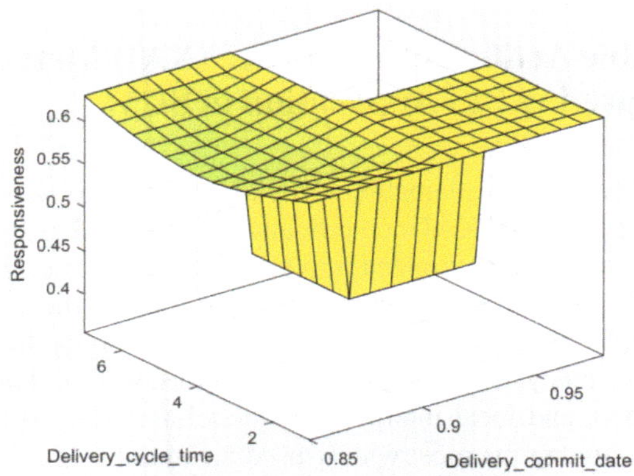

(Effects of "delivery commit date" and "delivery cycle time" on responsiveness)

Fig. 4.10 Input effect and interaction evaluation using RSM

As shown in Fig. 4.12, supply chain responsiveness is inversely proportional to the values of each input variable, which aligns with the expectations of supply chain managers. However, the FIRs in the FIS do not perfectly conform to this property, leading to some fluctuations and local violations in the PD curve. Even so, the trend remains correct.

Two-way PDPs can visualize the effects of two inputs at a time and evaluate their interaction.

$$PD(x_{m_1} = \xi_1, x_{m_2} = \xi_2) = E\big(COG(\tilde{A})\big(x_{m_1} = \xi_1, x_{m_2} = \xi_2\big)\big) \tag{4.8}$$

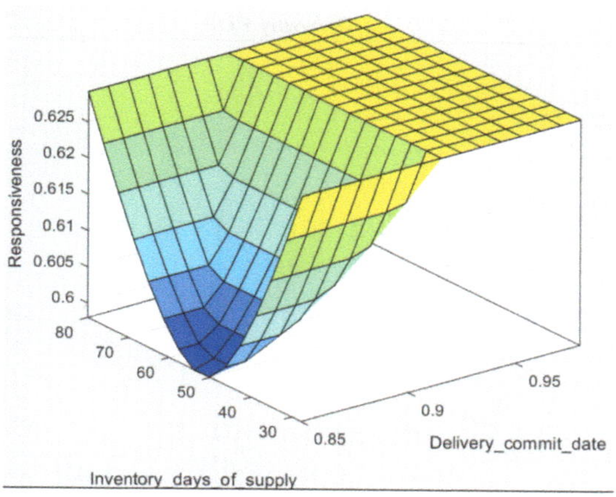

(Effects of "delivery commit date" and "inventory days of supply" on responsiveness)

Fig. 4.10 (continued)

```
% The PDs of 20 values are derived
PD1=zeros(20,1);
for l=1:20
   PDsum=0;
   % Average the results of 50 replicates as the expectation value
   for t=1:50
      % Derive the PD of "delivery commit date"
      i=0.85+(l-1)*0.0065;
      % Randomize the values of the other input variables
      j=27.6+rand*(80-27.6);
      k=1+rand*(7-1);
      PDsum=PDsum+evalfis([i j k],FIS1);
   end
   PD1(l,1)=PDsum/50;
end
x=1:20;
plot(x,PD1);
xlabel('Delivery commit date');
ylabel('Responsiveness');
title('One-way PDP');
```

Fig. 4.11 MATLAB program for deriving one-way PD

Example 4.3 For Example 4.1, a two-way PDP is drawn for each two input variables, for which the built-in GUI of MATLAB is applied, as illustrated in Fig. 4.13. The results are summarized in Fig. 4.14.

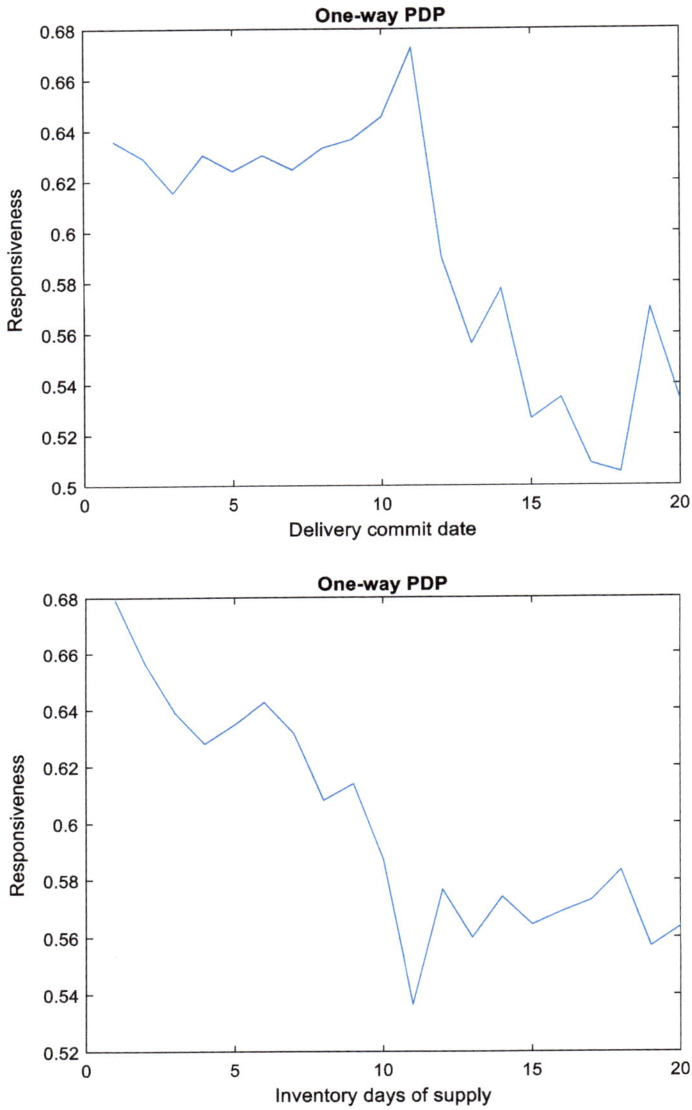

Fig. 4.12 One-way PDPs

4.5.2 SHAP Plots

SHAP analysis has been widely applied to evaluate the importance of each input variable in a black-box AI application [22]. SHAP analysis has also been used to

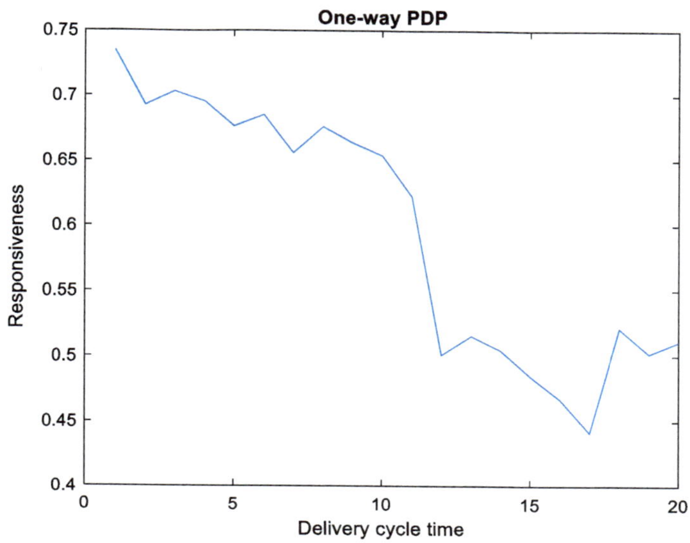

Fig. 4.12 (continued)

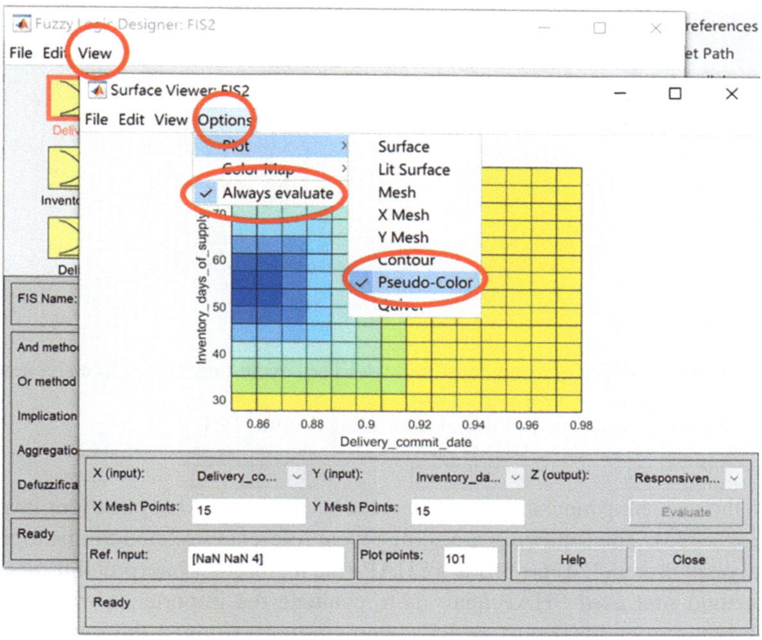

Fig. 4.13 Drawing two-way PDPs using the built-in GUI of MATLAB

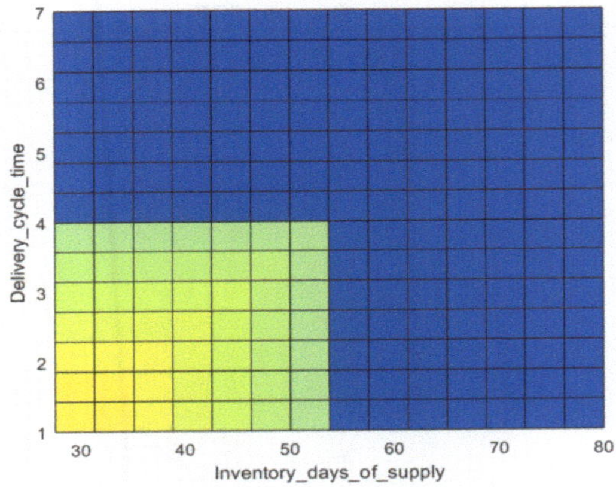

(Effects of "delivery cycle time" and "inventory days of supply" on responsiveness)

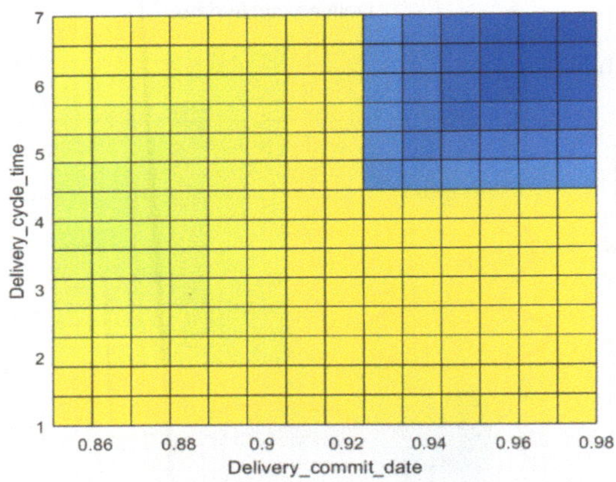

(Effects of "delivery commit date" and "delivery cycle time" on responsiveness)

Fig. 4.14 Input effect and interaction evaluation using two-way PDP

explain the reasoning mechanisms and/or results of various AI applications involving fuzzy logic [23]. The following are some recent references related to this topic.

Khalyasmaa et al. [24] proposed a SHAP-rule method for interpretable diagnosis. This method first used SHAP analysis to evaluate the importance of features for a black-box model (XGBoost or RF), then fuzzy partitioned the most important attributes and expressed them in linguistic terms. Finally, these linguistic terms were used to construct FIRs, ultimately forming an FIS.

Khan et al. [25] constructed a fuzzy deep neural network (FDNN) to predict epileptic seizures using electroencephalogram (EEG) recordings. From the FDNN,

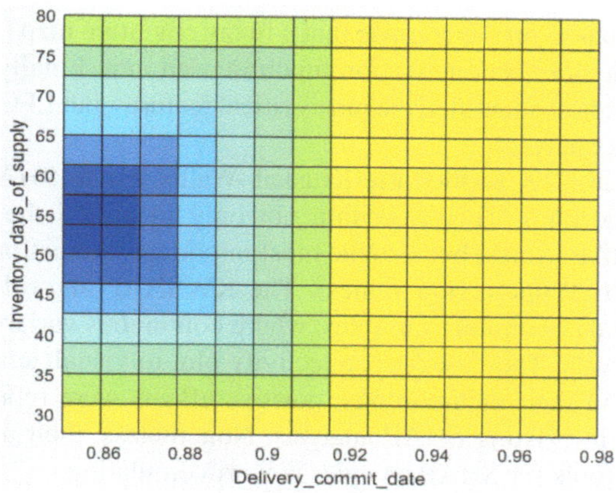

(Effects of "delivery commit date" and "inventory days of supply" on responsiveness)

Fig. 4.14 (continued)

some interpretable fuzzy inference rules could be derived, so the need to explain the inference mechanism was low. In addition, they also employed SHAP and LIME to identify the most important features for the prediction result. It was worth noting that since EEG was recorded continuously, these features were usually the average, peak, or other statistical terms of the original features.

To control the top temperature and pressure of an atmospheric tower, Gong et al. [26] combined a generative adversarial network (GAN) with a type-II fuzzy neural network, in which the GAN was used for data preprocessing, and the type-II fuzzy neural network was constructed for steady-state inverse prediction. They also conducted SHAP analysis to analyze and compare the effects of control parameters on the top temperature and pressure of the methanol distillation atmospheric tower.

Cao et al. [27] believed that the linguistic representation used by fuzzy inference rules expresses the knowledge or experience of experts in the form of human-like reasoning processes, which may not be easy to quantify, even if the reasoning process itself may be nonlinear and uncertain. Fuzzy inference rules are also easy to transmit or store. Therefore, they believed that fuzzy inference rules are an effective interpretation technology besides SHAP, LIME, decision rules, and heat maps. In addition, as an XAI technique, fuzzy inference rules have low algorithmic complexity, high transparency, and satisfactory expressive power, especially for tabular data.

Mansouri and Vadera [28] used a long short-term memory (LSTM) network based on vibration sensor readings to predict industrial bearing failures. They then built a learning fuzzy cognitive map (LFCM) to mimic the LSTM performance. Through static and dynamic analysis, they demonstrated that the LFCM could show which members of the reading sequence contributed to the prediction result and which values could be controlled to prevent possible failures. However, the interpretability of the LFCM was based on the premise that a fuzzy cognitive map (FCM) was interpretable despite the nonlinear activation function used in it. Finally, they compared LFCM with SHAP; both identified the most critical features, but LFCM could support additional network analyses.

Pérez-Sechi et al. [29] applied the Kruskal–Wallis test to SHAP and proposed the KWSHAP algorithm. This algorithm not only identified the most important features (using SHAP values) but also the most important interactions between these features (using SHAP interaction values). The KWSHAP algorithm could handle both categorical and continuous variables, where continuous variables were divided into multiple intervals. They used a connectivity plot to visualize the identification results. However, in their methodology, positive effects were represented in blue, rather than red as in existing SHAP analysis. Nonetheless, their attempt to design new visualization tools for SHAP analysis is worth emulating.

SHAP analysis is a commonly applied XAI technique to evaluate the impact (or importance) of each attribute on the model output. The SHAP value is defined as the weighted average of the marginal contributions of all possible coalitions $|\mathbf{F}|!$ as

$$\varphi_r(f) = \sum_{\{\mathbf{S} \subseteq \mathbf{F}\} \setminus \{r\}} \frac{|\mathbf{S}|!(|\mathbf{F}| - |\mathbf{S}| - 1)!}{|\mathbf{F}|!} \circ \left[f\left(x_{\mathbf{S} \cup \{r\}}\right) - f\left(x_{\mathbf{S}}\right) \right] \tag{4.9}$$

where

$\varphi_r(f)$ is the weighted average Shapley value that feature r provides in the context of all coalitions that exclude feature r;

$\mathbf{F}$ is the set of all features;

$\mathbf{S}$ is a subset (i.e., coalition) of $\mathbf{F}$;

$f\left(x_{\mathbf{S} \cup \{r\}}\right)$ is the model output considering feature r, while $f(x_{\mathbf{S}})$ is the model prediction/output without considering feature r.

A special case is when $|\mathbf{S}| = |\mathbf{F}| - 1$, i.e., $\mathbf{S} = \mathbf{F} \setminus \{r\}$,

$$\varphi_r(f) = \sum_{\{\mathbf{S} \subseteq \mathbf{F}\} \setminus \{r\}} \frac{|\mathbf{S}|!0!}{|\mathbf{F}|!} \circ \left[f\left(x_{\mathbf{S} \cup \{r\}}\right) - f\left(x_{\mathbf{S}}\right) \right] = \sum_{\{\mathbf{S} \subseteq \mathbf{F}\} \setminus \{r\}} \frac{\left[f\left(x_{\mathbf{S} \cup \{r\}}\right) - f\left(x_{\mathbf{S}}\right) \right]}{|\mathbf{F}|!} \tag{4.10}$$

In short, the SHAP value of an attribute refers to the average improvement in performance compared to a fully randomized baseline, given that the value of the attribute is fixed at its current value while other attributes are randomized [30].

One reason for the popularity of SHAP analysis is the visualization plots it offers to facilitate user interpretation, including force plots, waterfall plots, beeswarm plots, and scatter plots, which are supported by many software such as MATLAB, Python, R, and Excel. In these SHAP plots, positive and negative SHAP values are usually shown in red and blue, respectively, and the SHAP values of two adjacent attributes are connected to facilitate user interpretation.

Example 4.4 SHAP analysis is performed to evaluate and compare the importance of input variables to the supply chain responsiveness using the FIS. The required Python code is provided in Fig. 4.15. Take the following as an example:

FIS inputs: Delivery commit date = 92.5%; inventory days of supply = 52.5 days; delivery cycle time = 5.8 days.

FIS output: Supply chain responsiveness = 0.39.

The force plot and waterfall plot for this example are shown in Figs. 4.16 and 4.17, respectively, showing that delivery cycle time is the most important input variable for this example. In addition, the SHAP values of all input variables are negative, showing that the supply chain responsiveness decreases with the values of these input variables.

```
import xgboost
import shap
import pandas as pd
import numpy as np

# Import data generated by the FIS
FIS_data = pd.read_excel('F:\\papers\\SCM1\\FIS_data.xlsx')

X=FIS_data.iloc[0:1000,1:3]
y=FIS_data.iloc[0:1000,3]
y=y.to_numpy()

# Use XGBoost as a surrogate model
model1 = xgboost.XGBRegressor().fit(X, y)
O = model1.predict(X)

explainer1 = shap.Explainer(model1)
shap_values1 = explainer1(X)

# Compare the SHAP values of various input variables and draw plots
shap.plots.waterfall(shap_values1[7])
shap.force_plot(shap_values1[7], matplotlib=True)
shap.plots.scatter(shap_values1[:, "Delivery commit date"], color=shap_values1)
shap.plots.beeswarm(shap_values1)
```

Fig. 4.15 Python code for performing SHAP analysis

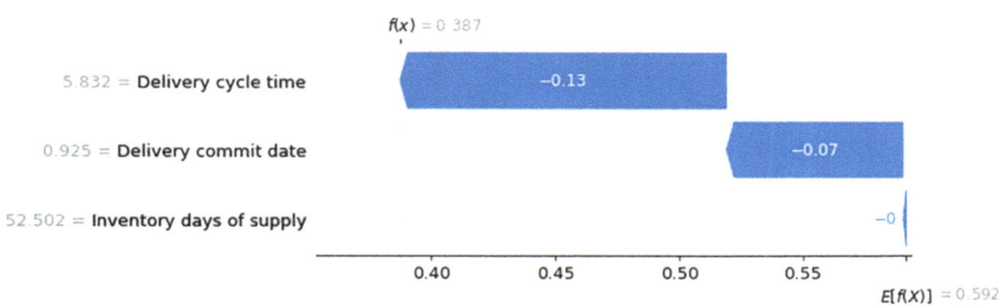

Fig. 4.16 Waterfall plot for the example

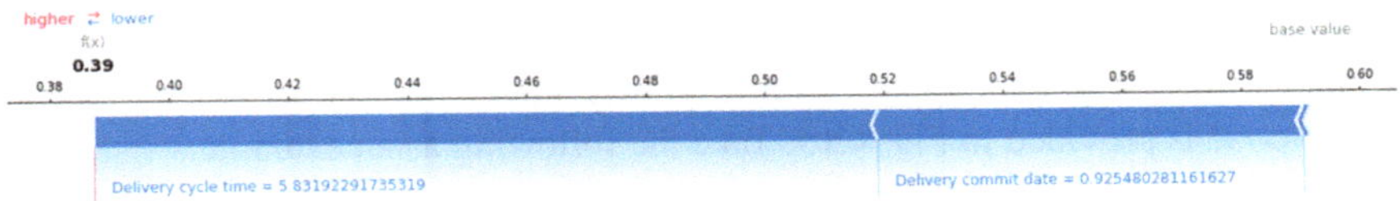

Fig. 4.17 Force plot for the example

The beeswarm plot considering all examples is shown in Fig. 4.18, revealing that delivery cycle time is also the most important input variable for all examples. To further investigate its effect, a scatter plot is drawn in Fig. 4.19, showing that the attribute having the highest correlation with delivery cycle time is delivery commit date. When delivery cycle time is less than 4, the SHAP value of delivery cycle time increases with the value of delivery commit date. In contrast, when delivery cycle time is greater than 4, the SHAP value of delivery cycle time decreases with the value of delivery commit date.

Example 4.5 Consider an FIS comprising the following three FIRs:

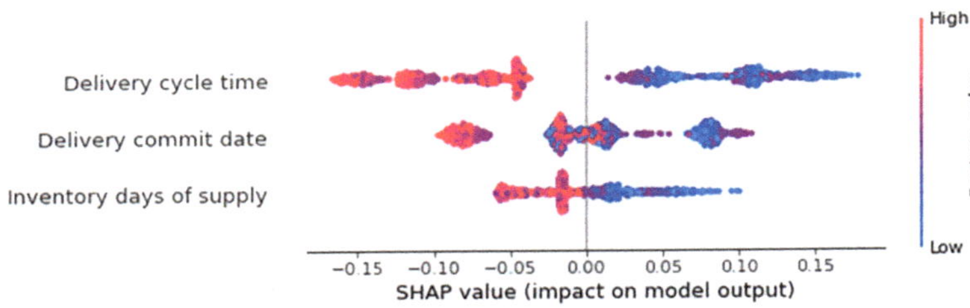

Fig. 4.18 Beeswarm plot considering all examples

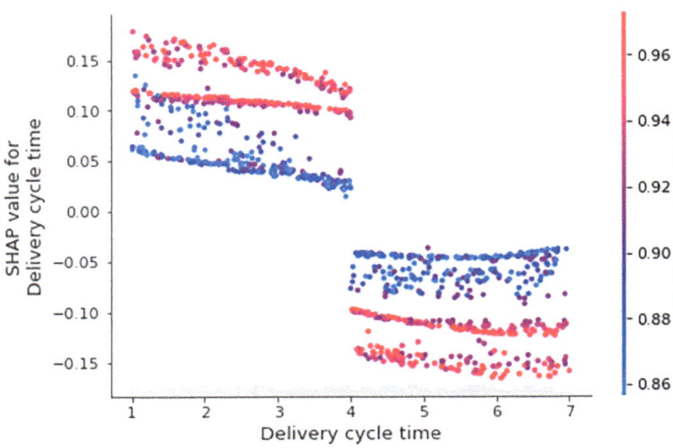

Fig. 4.19 Scatter plot for delivery cycle time

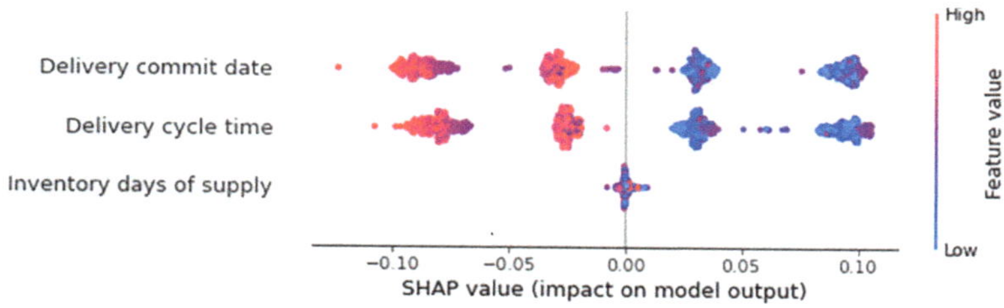

Fig. 4.20 Beeswarm plot for the case in Example 2.5

IF Delivery commit date is High AND Delivery cycle time is High THEN Responsiveness is Low.

IF Delivery commit date is Low AND Inventory days of supply is Medium AND Delivery cycle time is Medium THEN Responsiveness is Medium.

IF Delivery commit date is Medium AND Delivery cycle time is Low THEN Responsiveness is Medium.

The beeswarm plot is drawn in Fig. 4.20. Obviously, delivery commit date is the most important input variable affecting supply chain responsiveness, followed by delivery cycle time. The least important input variable is inventory days of supply. The order of importance levels of these input variables is consistent with their frequency of occurrence.

The following conclusions are derived:

- Similar to decision rules, input variables not appearing in an FIR are not important. Input variables appearing in the same FIR have unequal importance levels.
- The input variables that appear in more FIRs are likely to be more important.
- Furthermore, in random forests (RF), the importance levels of decision trees vary even though averaging is used to aggregate their results. Similarly, the importance levels of the FIRs in an FIS also vary. For example, if the output value of one FIR is contained in the output value of another FIR, then the former FIR becomes unimportant because FU is applied to aggregate the outputs of all FIRs.

4.5.3 RF-Based Incremental Interpretation (RFII)

Linguistic terms are widely used in interpersonal communication, so an FIR, which consists of linguistic terms, is inherently interpretable and does not require interpretations. However, similar to the problem with RFs, multiple FIRs can be applied to a single case simultaneously, leading to confusion in the communication of the inference result. Such a problem can be solved using a methodology similar to random forest-best incremental interpretation (RFII) [31] as follows:

Step 1. Aggregate the outputs of all FIRs as $\hat{o}_j$ for example j.

Step 2. Set $q = 1$.

Step 3. Choose the FIR with the output closest to $\hat{o}_j$.

Step 4. Set the q-th incremental rule to the FIR.
Step 5. Set $\hat{o}_j(current)$ to the output of the FIR.
Step 6. Remove the FIR.
Step 7. Increase q by 1.
Step 8. If $q > Q$ (the number of FIRs), go to Step 14; otherwise, go to Step 9.
Step 9. Consider one more FIR. Find the set with the output closest to $\hat{o}_j$.
Step 10. Subtract $\hat{o}_j(current)$ from the output of the set: The result is indicated by $\Delta\hat{o}_j(current)$.
Step 11. Set the premise of the q-th incremental rule to that of the additional FIR.
Step 12. Set the output of the q-th incremental rule to "add $\Delta\hat{o}_j(current)$ to $\hat{o}_j$" or "subtract $-\Delta\hat{o}_j(current)$ from $\hat{o}_j$", and "so the output becomes the output of the set".
Step 13. Return to Step 5.
Step 14. End.
This methodology is named FIS-based incremental interpretation (FISII).

Example 4.6 An FIS composed of the following three FIRs:

IF Delivery commit date is High AND Inventory days of supply is Low AND Delivery cycle time is Low THEN Responsiveness is Low.

IF Delivery commit date is Medium AND Inventory days of supply is Medium AND Delivery cycle time is Low THEN Responsiveness is Medium.

IF Inventory days of supply is Medium AND Delivery cycle time is Medium THEN Responsiveness is Medium.

Consider the following example: delivery commit date = 93%, inventory days of supply = 45 days, and delivery cycle time = 3.5 days. The supply chain responsiveness is obtained as 0.39 (see Fig. 4.21). All FIRs are applicable to this example, i.e., membership >0, which causes user confusion. To address this, FIRII is applied as follows.

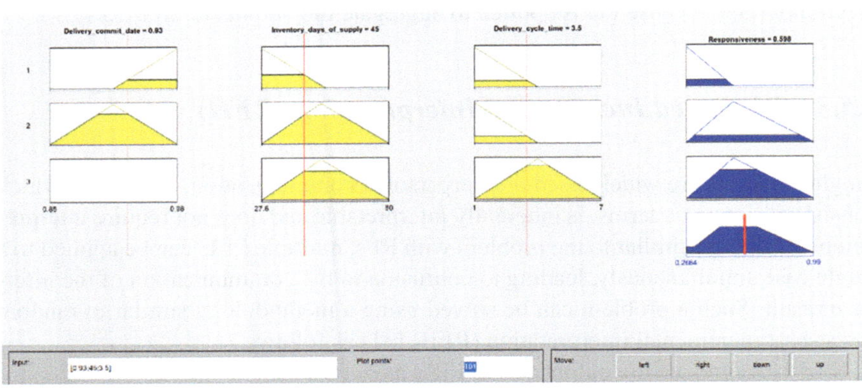

Fig. 4.21 Fuzzy inference result for Example 4.6

Step 1. $\hat{o}_j = 0.598$.

Step 2. $q = 1$.

Step 3. The output (COG) of each FIR is presented in Table 4.5. Therefore, the output of FIR #3 is closest to $\hat{o}_j$.

Step 4. The 1st incremental rule = FIR #3 = "IF Inventory days of supply is Medium AND Delivery cycle time is Medium THEN Responsiveness is 0.602 (Medium)".

Step 5. $\hat{o}_j(current) = 0.602$.

Step 6. Remove FIR #3.

Step 7. $q = 2$.

Step 8. $q < Q = 3$, go to Step 9.

Step 9. By additionally considering FIR #1, the output is closest to $\hat{o}_j$ (see Table 4.6).

Step 10. $\Delta\hat{o}_j(current) = 0.598 - 0.602 = -0.004$.

Step 11. The premise of the 2nd incremental rule = the premise of FIR #1 = "IF Delivery commit date is High AND Inventory days of supply is Low AND Delivery cycle time is Low".

Step 12. The output of the 2nd incremental rule = "Subtract Responsiveness by 0.004, so Responsiveness becomes 0.598".

Step 13. Return to Step 6.

Step 6. Remove FIR #1.

Step 7. $q = 3$.

Step 8. $q = Q = 3$, go to Step 9.

Step 9. By additionally considering FIR #2, the output is closest to $\hat{o}_j$ (see Table 4.7).

Step 10. $\Delta\hat{o}_j(current) = 0.598 - 0.597 = 0$.

Step 11. The premise of the 3rd incremental rule = the premise of FIR #2 = "IF Delivery commit date is Medium AND Inventory days of supply is Medium AND Delivery cycle time is Low".

Step 12. The output of the 3rd incremental rule = "Add 0 to Responsiveness, so Responsiveness becomes 0.598".

Step 13. Return to Step 6.

Table 4.5 Output of each FIR

FIR #	Output
1	0.389
2	0.622
3	0.602

Table 4.6 Considering one more FIR

FIR #	Output
3 & 1	0.598
3 & 2	0.602

Table 4.7 Considering the third FIR

FIR #	Output
3 & 1 & 2	0.598

Table 4.8 Comparison of the original result and the FISII interpretation

Method	Interpretation
Original result	This example satisfies the following rules "IF Inventory days of supply is Medium AND Delivery cycle time is Medium THEN Responsiveness is Medium" "IF Delivery commit date is High AND Inventory days of supply is Low AND Delivery cycle time is Low THEN Responsiveness is Low" "IF Delivery commit date is Medium AND Inventory days of supply is Medium AND Delivery cycle time is Low THEN Responsiveness is Medium" So, the supply chain responsiveness is evaluated as 0.598
FISII interpretation	This example satisfies the following rules "IF Inventory days of supply is Medium AND Delivery cycle time is Medium THEN Responsiveness is 0.602 (Medium)" "IF Delivery commit date is High AND Inventory days of supply is Low AND Delivery cycle time is Low THEN Subtract Responsiveness by 0.004, so Responsiveness becomes 0.598" "IF Delivery commit date is Medium AND Inventory days of supply is Medium AND Delivery cycle time is Low THEN Add 0 to Responsiveness, so Responsiveness becomes 0.598" So, the supply chain responsiveness is evaluated as 0.598

Step 6. Remove FIR #2.
Step 7. $q = 4$.
Step 8. $q > Q = 3$, go to Step 14.
Step 14. End.

The first incremental rule tells the user that the evaluated supply chain responsiveness is around 0.602 (medium). This rule is the most important rule. The user can stop reading if he/she doesn't want to continue reading. The user reads the rest of the rules from top to bottom, revising the evaluated supply chain responsiveness successively, and finally learns that the evaluated supply chain responsiveness is 0.598 h.

A comparison of the original explanation and the FISII interpretation is provided in Table 4.8.

4.5.4 LIME

LIME is to fit the local feasible region around an example with a simpler model (the approximate AI application) such as linear regression (LR) or a decision tree (DT), so as to avoid overfitting and enhance transparency. The steps of implementing LIME include.

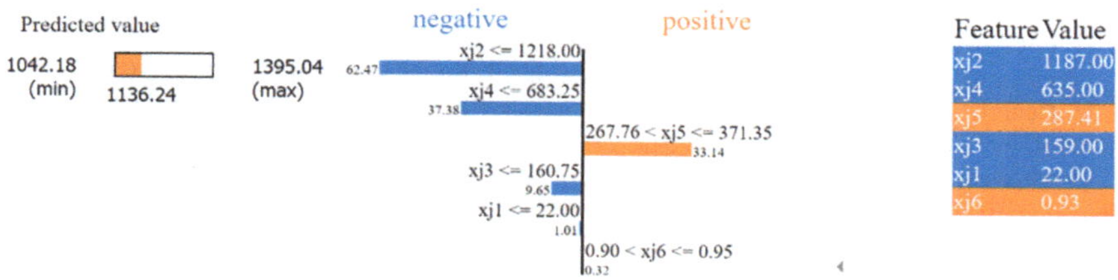

Fig. 4.22 LIME summary plot

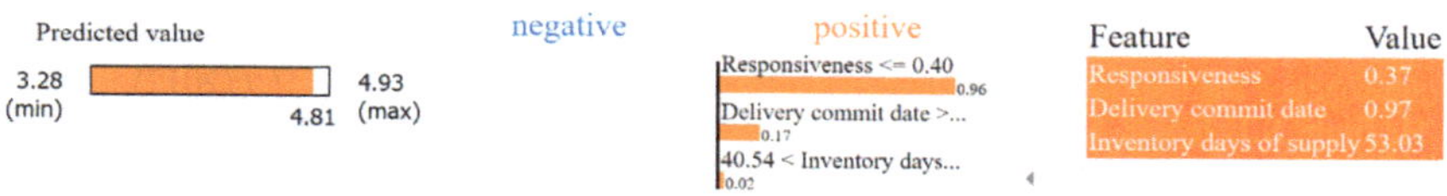

Fig. 4.23 LIME summary plot for Example 4.6

- Pre-model data analysis.
- AI model training.
- Synthetic data generation.
- AI application to synthetic data.
- Approximate AI application (fit synthetic data).
- Explanation performance evaluation.

In a LIME summary plot (see Fig. 4.22), the output (or prediction) of an example relative to the output range is shown. In addition, the value of each feature is also provided. The contribution of the feature on the output is also evaluated based on the simpler model. Similar to a waterfall model in SHAP, positive and negative contributions are shown in different colors.

Example 4.7 For the case depicted in Example 4.6, a LIME summary plot is drawn, as shown in Fig. 4.23.

References

1. B. Fahimnia, C.S. Tang, H. Davarzani, J. Sarkis, Quantitative models for managing supply chain risks: a review. Eur. J. Oper. Res. **247**(1), 1–15 (2015)
2. T.C.T. Chen, Y.C. Wang, H.C. Wu, An explainable fuzzy collaborative intelligence approach for prioritizing possible measures to tackle challenges in semiconductor manufacturing supply chain localization. Oper. Res. Int. J. **26**(1), 7 (2026)
3. T.C.T. Chen, Localization of semiconductor supply chains—Driving forces and challenges, in *Supply Chain Localization in the Semiconductor Industry: Rebuilding the Competitiveness and Sustainability of Semiconductor Manufacturers* (2025), pp. 23–44
4. B.R. Tukamuhabwa, M. Stevenson, J. Busby, M. Zorzini, Supply chain resilience: definition, review and theoretical foundations for further study. Int. J. Prod. Res. **53**(18), 5592–5623 (2015)

5. E. Pourjavad, A. Shahin, The application of Mamdani fuzzy inference system in evaluating green supply chain management performance. Int. J. Fuzzy Syst. **20**(3), 901–912 (2018)
6. H. Behret, B. Öztayşi, C. Kahraman, A fuzzy inference system for supply chain risk management, in *Practical Applications of Intelligent Systems: Proceedings of the Sixth International Conference on Intelligent Systems and Knowledge Engineering* (2011), pp. 429–438
7. R. Pabarja, G. Jamali, K. Salimifard, A. Ghorbanpur, Analysis of the LARG of the hospital medical equipment supply chain using the fuzzy inference system. Int. J. Res. Indus. Eng. **13**(2), 116–151 (2024)
8. G.M.D. Ganga, L.C.R. Carpinetti, A fuzzy logic approach to supply chain performance management. Int. J. Prod. Econ. **134**(1), 177–187 (2011)
9. Association for Supply Chain Management, Introduction to processes. https://scor.ascm.org/processes
10. T. Efendigil, S. Önüt, An integration methodology based on fuzzy inference systems and neural approaches for multi-stage supply-chains. Comput. Ind. Eng. **62**(2), 554–569 (2012)
11. H. Jung, S.J. Jeong, Managing demand uncertainty through fuzzy inference in supply chain planning. Int. J. Prod. Res. **50**(19), 5415–5429 (2012)
12. M. Momenitabar, Z.D. Ebrahimi, M. Arani, J. Mattson, P. Ghasemi, Designing a sustainable closed-loop supply chain network considering lateral resupply and backup suppliers using fuzzy inference system. Environ. Dev. Sustain. **27**, 23203–23236 (2025)
13. A. Amindoust, A. Saghafinia, Textile supplier selection in sustainable supply chain using a modular fuzzy inference system model. J. Textile Inst. **108**(7), 1250–1258 (2017)
14. T. Chen, A fuzzy ubiquitous traveler clustering and hotel recommendation system by differentiating travelers' decision-making behaviors. Appl. Soft Comput. **96**, 106585 (2020)
15. L. Pokorádi, B. Szamosi, Fuzzy failure modes and effects analysis with summarized center of gravity defuzzification, in *2015 16th IEEE International Symposium on Computational Intelligence and Informatics* (2015), pp. 147–150
16. X. Liu, Parameterized defuzzification with maximum entropy weighting function—another view of the weighting function expectation method. Math. Comput. Model. **45**(1–2), 177–188 (2007)
17. D. García-Zamora, A.F.R.L.D. Hierro, H. Bustince, A note on the admissibility of the centroid-based preorder for fuzzy numbers. Comput. Appl. Math. **45**(4), 150 (2026)
18. T.C.T. Chen, XAmI applications to smart homes, in *Explainable Ambient Intelligence (XAmI) Explainable Artificial Intelligence Applications in Smart Life* (2024), pp. 39–61
19. U. Kamath, J. Liu, *Explainable Artificial Intelligence: An Introduction to Interpretable Machine Learning* (Springer, 2021)
20. C. So, Understanding the prediction mechanism of sentiments by XAI visualization, in *Proceedings of the 4th International Conference on Natural Language Processing and Information Retrieval* (2020), pp. 75–80
21. Y.C. Wang, T. Chen, Adapted techniques of explainable artificial intelligence for explaining genetic algorithms on the example of job scheduling. Expert Syst. Appl. **237**(A), 121369 (2024)
22. Y. Meng, N. Yang, Z. Qian, G. Zhang, What makes an online review more helpful: an interpretation framework using XGBoost and SHAP values. J. Theor. Appl. Electron. Commer. Res. **16**(3), 466–490 (2020)
23. Y.C. Lin, T.C.T. Chen, An intelligent system for assisting personalized COVID-19 vaccination location selection: Taiwan as an example. Digit. Health **8**, 20552076221109064 (2022)
24. A.I. Khalyasmaa, P.V. Matrenin, S.A. Eroshenko, Interpretable diagnostics with SHAP-Rule: fuzzy linguistic explanations from SHAP values. Mathematics **13**(20), 3355 (2025)
25. F.A. Khan, Z. Umar, A. Jolfaei, M. Tariq, Explainable fuzzy deep learning for prediction of epileptic seizures using EEG. IEEE Trans. Fuzzy Syst. **32**(10), 5428–5437 (2024)
26. Y. Gong, Q. Zhang, Y. Ren, Z. Liu, M.T. Abu Seman, Research on fuzzy control of methanol distillation based on SHAP (SHapley Additive exPlanations) interpretability and generative artificial intelligence. Sensors **25**(5), 1308 (2025)
27. J. Cao, T. Zhou, S. Zhi, S. Lam, G. Ren, Y. Zhang, et al., Fuzzy inference system with interpretable fuzzy rules: advancing explainable artificial intelligence for disease diagnosis—a comprehensive review. Inf. Sci. **662**, 120212 (2024)

28. T. Mansouri, S. Vadera, Explainable fault prediction using learning fuzzy cognitive maps. Expert. Syst. **40**(8), e13316 (2023)
29. C.I. Pérez-Sechi, I. Gutiérrez, J. Castro, D. Gómez, D. Martín, R. Espínola, On the role of relations and interactions in fuzzy systems for machine learning, in *2025 IEEE International Conference on Fuzzy Systems* (2025), pp. 1–6
30. W.E. Marcílio, D.M. Eler, From explanations to feature selection: assessing SHAP values as feature selection mechanism, in *2020 33rd SIBGRAPI Conference on Graphics, Patterns and Images* (2020), pp. 340–347
31. T. Chen, Y.C. Wang, A two-stage explainable artificial intelligence approach for classification-based job cycle time prediction. Int. J. Adv. Manuf. Technol. **123**(5), 2031–2042 (2022)

MIX
Papier aus verantwortungsvollen Quellen
Paper from responsible sources
FSC® C105338

If you have any concerns about our products,
you can contact us on
ProductSafety@springernature.com

In case Publisher is established outside the EU,
the EU authorized representative is:
Springer Nature Customer Service Center GmbH
Europaplatz 3, 69115 Heidelberg, Germany

Printed by Libri Plureos GmbH
in Hamburg, Germany